CROP PRODUCTION SCIENCE IN HORTICULTURE SERIES

Series Editors: Jeff Atherton, Senior Lecturer in Horticulture, University of Nottingham, and Alun Rees, Horticultural Consultant and Editor, *Journal of Horticultural Science*

This series examines economically important horticultural crops selected from the major production systems in temperate, sub-tropical and tropical climatic areas. Systems represented range from open field and plantation sites to protected plastic and glass houses, growing rooms and laboratories. Emphasis is placed on the scientific principles underlying crop production practices rather than on providing empirical recipes for uncritical acceptance. Scientific understanding provides the key to both reasoned choice of practice and the solution of future problems.

Students and staff at universities and colleges throughout the world involved in courses in horticulture, as well as in agriculture, plant science, food science and applied biology at degree, diploma or certificate level will welcome this series as a succinct and readable source of information. The books will also be invaluable to progressive growers, advisors and end-product users requiring an authoritative, but brief, scientific introduction to particular crops or systems. Keen gardeners wishing to understand the scientific basis of recommended practices will also find the series very useful.

The authors are all internationally renowned experts with extensive experience of their subjects. Each volume follows a common format covering all aspects of production, from background physiology and breeding, to propagation and planting, through husbandry and crop protection, to harvesting, handling and storage. Selective references are included to direct the reader to further information on specific topics.

Titles in Preparation:
Onions and other Vegetable Alliums J. L. Brewster
Cucurbits R. W. Robinson
Citrus F. S. Davies and L. G. Albrigo
Tomatoes J. Atherton
Ornamental Bedding Plants A. M. Armitage
Apples and Pears W. R. Autio
Carrots and Related Vegetable Umbelliferae V. E. Rubatzky, P. W. Simon and C. F. Quiros

If of thy wordly goods thou art bereft,
And of thy slender store there be but left
Two loaves, sell one, and with the dole,
Buy daffodils to feed thy soul.

For Mo, David and Siân

Ornamental Bulbs, Corms and Tubers

A.R. Rees

C·A·B International

C·A·B International   Tel: Wallingford (0491) 32111
Wallingford   Telex: 847964 (COMAGG G)
Oxon OX10 8DE   Telecom Gold/Dialcom: 84: CAU001
UK   Fax: (0491) 33508

A catalogue record for this book is available from the British Library

ISBN 0 85198 656 0

Typeset by Alden Multimedia Ltd
Printed and bound in the UK by Redwood Press Ltd, Melksham

CONTENTS

PREFACE

A major problem in writing this book was to condense an enormous amount of information into a small volume, whilst attempting to preserve sufficient coherence and readability to retain the reader's interest and provide entry points to other, more detailed, literature on specific items. Wherever possible, general principles have been stressed, and common examples have been considered in some detail, with less information on other species.

The sector of the horticultural industry concerned with 'bulb' plants is involved with producing 'bulbs' for sale (the dry bulb trade), with producing cut flowers for sale, both natural season and out-of-season (forced) and with pot-plant production. Specialists breed, select and propagate new cultivars, and 'bulbs' are used for amenity and decorative purposes in parks and gardens. There is scientific and advisory back-up to deal with problems of plant health, methods of storing and mechanically handling 'bulbs', in bulk quantities, and for devising new schedules of temperature treatments for 'bulb' storage, for accurate control of the timing of flowering and for maintaining flower quality. For all these activities, a background knowledge of the physiology of the plant is essential.

The subject is vast: of the ornamental geophytes, taking both terms in the widest sense, there are tens of families, hundreds of genera, perhaps thousands of species, and certainly tens of thousands of cultivars. Of these, few species are commercially important and little is known about most of the minor ones, except for recorded observations on hardiness, planting times and flowering dates in outdoor situations, i.e. gardening lore. In contrast, detailed information and recommendation down to cultivar level, based on years of research, development and experience, are available for those of major commercial significance.

Horticulturally, these plants can be grouped, rather artificially, into many categories: hardy, spring flowering plants, less hardy summer and autumn flowering plants, those forced into flower out of their natural season, or grown for cut-flower production under glass, or naturalized for amenity use, and tender house plants, often of tropical or sub-tropical origin, grown as much for their foliage as their flowers.

The storage organs themselves are of several kinds, morphologically and physiologically, and flowers are produced at different times, seasonally, and relative to the plant's periodicity and the development of other organs, being synanthous or hysteranthous. Geophytes can be evergreen or deciduous, annual or perennial, and can even change their growth and flowering characteristics in cultivation or in another environment, as in the case of *Ornithogalum thyrsoides*, a deciduous perennial which becomes ever-flowering in a tropical environment, exhibiting no dormancy, and producing no bulbs (Halevy, 1990). Nearly all geophytes exhibit some 'dormancy', again of several kinds and at different times of year, which can be broken in different ways.

Over the past 20 years there have been five International Symposia on flower bulbs and allied crops, organized by the International Society for Horticultural Science, the sixth was held in 1992. Many other national and international meetings deal mainly or partially with bulbs, and there are many books intended for amateur gardeners, commercial growers and research scientists. It is estimated that over 5000 scientific papers have been written on these plants, and over 300 abstracts on ornamental bulbs are added to the CAB ABSTRACTS database each year. Writing this book therefore seemed like an attempt to linearize an enormous multi-dimensional matrix containing large lacunae.

Inevitably, there have been problems with nomenclature, because the book aims to serve science and horticulture, each with its own jargon. In two cases I have had to invent my own rules. Dormancy to a horticulturist is a condition when a plant is not growing, but to a plant scientist it represents a more fundamental state when a meristem is inactive, for whatever reason. To the former, a tulip bulb in store after lifting in July is dormant, but the scientist knows that behind that quiescent exterior, leaf and flower differentiation are in active progress, so the bulb is not really dormant. For the general and horticultural usage, I have therefore employed 'dormant', omitting the inverted commas for more fundamental consideration of meristems. Similarly, the term 'bulb' (within inverted commas) is used in its horticultural sense for all the storage organs in the title of this book, and a few more. By this means I avoid a string of names every time I refer to storage organs in general. I hope my botanical friends will understand.

For critical and constructive comments on this book, I am most grateful to Gordon Hanks, my former colleague and research collaborator at the former Glasshouse Crops Research Institute in the heyday of UK bulb research. The responsibility for such errors as remain is wholly mine.

Alun Rees
Rustington
West Sussex UK, 1992

INTRODUCTION

Plants with storage organs are found in all the climatic areas of the world, from a number of botanical families, and, as the title of the book suggests, with a range of types of storage organ. Their unity as a group rests on the possession of a storage organ which can provide reserves to allow the plant to survive the unfavourable periods of its native habitat, and to grow and flower at times of year when development cannot easily be supported by current synthesis of food materials by aerial parts.

ECOLOGICAL AND HORTICULTURAL IMPLICATIONS

Such characteristics have obvious ecological advantages allowing the plant to survive in a quiescent mode these unfavourable periods (whether too dry, too hot or too cold) until conditions improve. Storage organs resemble seeds in containing the plant in miniature, or the means of generating a new plant. They are conveniently packaged with a good supply of food reserve to maintain the plant's viability and allow development to continue despite a harsh external environment, protected from predators and from desiccation by one or more protective outer layers. The storage organs often act as a means of propagation, itself a survival mechanism, as when one plant produces, in a single season, several storage organs, each capable of independent existence. Plant dispersal mechanisms also operate through the storage organs, either intrinsically to the plant such as the production of storage organs at the ends of stolons or involving other means, such as contractile organs to separate the daughter propagules from the parent either laterally or vertically. Dispersal may also result from the activities of small mammals removing storage organs from the parent plant, or they may be carried by water.

    Possession of storage organs also allows the plant some independence from the periodicity of its habitat. Flowering can be achieved earlier in the year than in an annual which has to germinate and pass through a juvenile phase with a build-up of vegetative structure to support the demands of flowering and subsequent seed maturation. Flowering can be completely separated from above-ground vegetative development because of the availability of food reserves, whilst the initiation and early growth of daughter 'bulbs' can proceed underground at the expense of previously laid down reserves out of sight of grazing animals seeking to feed on fresh new growth.

Such advantages to the plant are also advantages in horticulture. 'Bulbs' are propagated naturally and are suitable for distribution and sale in trade, with a good storage life and resistance to deterioration, and come true-to-type because they are vegetatively produced and genetically the same as the mother plant. They can be manipulated to flower outside their natural flowering times because of their lack of dependence on external conditions, and their capacity to respond predictably to imposed conditions, especially temperature.

The importance of perennation was stressed by Raunkier (1934) whose universal system of plant life forms was based on the position of the buds which survive the unfavourable season, relative to the soil surface. His group most relevant to us is the cryptophytes, whose buds are buried in the ground (geophytes) or occur below water, in marshes, ponds and lakes (helophytes and hydrophytes). The geophytes are subdivided into those whose storage organs are rhizomes, bulbs, stem tubers and root tubers, well established names still in common use.

With increasing protection afforded to the perennating buds as they are, successively, just above the soil surface, at the soil surface or below it, there is a reduction in the plant mass which survives the unfavourable season, and therefore a greater need for more storage tissue to ensure survival. Further, as the better protection afforded to the surviving buds means that the plants can grow in areas where the unfavourable season is prolonged, there is a greater need to ensure rapid growth at the start of the short growing season. In many geophytes this is achieved by an active period of underground development ensuring that in spring, shoot growth is rapid, and flowering is synchronous – a feature well illustrated by the rich floral display at the start of the spring in the steppes. Such a growth strategy must be based on a large supply of reserve materials in the plants' storage organs. In the most extreme climates in near-desert conditions, where annual rainfall is extremely low, variable and undependable, after an exceptionally dry year plants do not flower or may not even produce aerial parts; survival then depends on sufficient reserves to last an 'extra' year.

Storage organs commonly occur in three situations: where the plant structure is considerably reduced for perennation, and must be replaced at least partially and initially from storage reserves, where the growing season is short and development occurs in the 'dormant' period, and where the plant development is phasic such that reserves are accumulated in one season to allow a massive development such as flowering and fruiting in another season (the following one in biennials, or considerably later in monocarpic perennials). These considerations apply to wild plants. In developing horticultural plants, man has accentuated the desired characteristics to his own needs; although many wild characters are retained in developing selected cultivars, direct comparisons with wild plants are not valid.

ORIGINS OF THE GEOPHYTIC HABIT

Plants with storage organs are not confined to Angiosperms as horsetails (*Equisetum*) and several ferns have rhizomes, whilst the occurrence of rhizomes in several of the earliest Devonian fossil plants indicates that the geophytic habit originated early in plant evolution, and probably on many occasions because of the different types of storage organs in many plant groups.

Consideration of the evolution of the geophytic habit has largely been restricted to the monocotyledons, where it is most common. There is a lack of unanimity about the ancestral monocotyledon, with some authorities favouring a reductionist theory involving the loss of a cambium by a shrublet ancestor. In contrast, the closely argued hypothesis of Holttum (1955) favours an ancestral monocotyledonous plant with no cambium, which led directly to the development of a continuous sympodial growth habit in the moist tropics where these plants are thought to have evolved. The greater species abundance of monocotyledonous plants in the moist tropics, their maximum development between latitudes 45°N and S, and their inability to withstand hard frost support this origin.

The absence of secondary thickening means that growth in shoot diameter is achieved by primary thickening of the narrow embryonic axis, a slow process in which many internodes are formed at or near ground level. This slow establishment growth is probably why annual monocotyledons are rare and biennials unknown. Both adventitious buds and roots are confined to the nodes, and despite means of improving the plant–soil contact by such devices as prop roots and contractile roots for pulling the plant into the soil, the form of the plant is constrained by the physical strength of the base of the stem and its limited conducting capacity. Branching of the above-ground parts is therefore restricted in physical extent and in life span. The plant grows by producing new adventitious shoots from the buds at the base of the plant; these have the same form as the parent shoot, with their own adventitious buds and roots; this growth pattern produces the range of readily identified tufted growth forms, based on sympodial growth, shown by bamboos, bananas, orchids, aroids and corm- and bulb-forming geophytes. The root system is also adventitious because of the radicle's initial small size and inability to grow radially, a limitation counteracted by the many short stem internodes near ground level allowing many roots to grow.

Even in a non-seasonal environment, growth of buds of potential new basal shoots would be constrained by apical dominance or correlative inhibition so that some would be inactive or 'dormant', growing only when the dominant shoot flowered, senesced or died naturally (deterministic growth) or succumbed to some predation (opportunistic growth). Such a system appears poised to respond to climatic change or a move to an area of seasonal climate. The required adaptations are that each new lateral shoot should arise from an underground bud, should remain as a bud or continue diageotropic growth (preferably below ground) and should develop storage tissue and reserves adequate to support the plant until favourable conditions return. For such a strategy to be successful requires several physiological and biochemical controls. There is a need to inhibit extension growth in favour of lateral growth, to develop plant processes to respond to environmental stimuli via hormones, to develop an outer covering to restrict water loss, to lower respiration rate to conserve existing food reserves, to protect the storage organ from frost if necessary, and to deter fungal and animal pests from consuming the reserves, which would be scarce in the season unfavourable for growth. All these requirements have been met in different ways in different plant groups and storage organs, but little is known about how they developed. In all, there is a common strategy of coping with extreme conditions by the vulnerable above-ground parts dying down completely in the unfavourable season(s), with growth being resumed by the underground survival parts.

Many of the features described above for monocotyledonous plants also occur in the dicotyledonous geophytes, such as sympodial growth, the absence of a cambium, basal growth from replacement buds and adventitious roots.

Little is known about the way storage organs evolved, despite their obvious importance in food plants. Storage tissues are usually parenchymatous, and often closely associated with phloem tissue which aids sink filling and emptying. In the evolution of a storage organ and in the tuberization process of an existing, growing organ, it is necessary that there is a preponderance of radial over longitudinal growth, usually first seen during development as a cessation of the latter, followed by a stimulation of lateral growth. Environmental signals such as temperature or day length often trigger the process, acting via growth regulators. The three successive steps are growth inhibition, tuberization and 'dormancy'.

There is a wide range of stored food materials; starch is the most usual, with glucose, sucrose and highly polymerized glucofructosans such as inulin occurring commonly. It is suspected that some of the more highly polymerized carbohydrates have a cryo-protective function, although they are also broken down into simpler forms and are metabolized. The reserves are protected by outer layers of tissue such as old leaf bases or scales which restrict water loss, whilst the presence of toxic, repellent or deterrent chemicals such as alkaloids, oxalic acid, antimitotic agents and the 'flavour' chemicals of the onion and garlic protect the reserves from predation by insects and microorganisms. Respiration rates of 'dormant' 'bulbs' are generally low, and in those species that have been examined (iris and tulip), are also unexpectedly low at high temperatures. The control mechanism for this is unknown.

OCCURRENCE BY FAMILY

Ornamental bulb and corm plants occur in a number of families of both monocotyledons and dicotyledons, but numerically, the vast majority are monocotyledonous and belong to a few families: within the order Asparagales, the Alliaceae, the Amaryllidaceae and Hyacinthaceae, and within the order Liliales, the Alstroemeriaceae, Iridaceae and Liliaceae. Geophytes also occur in many other monocotyledonous families, such as the Agavaceae, the Araceae, the Asphodelaceae, the Colchicaceae, the Convallariaceae, the Hemerocallidaceae and the Tecophilaeaceae.

According to Dahlgren and Clifford (1982) true bulbs are an almost exclusive feature of the Liliiflorae, and confined to the Liliales and Asparagales. The Liliaceae have many bulbous species, in contrast to the Iridaceae with few. In the Asparagales, most members of the Alliaceae, the Amaryllidaceae and Hyacinthaceae are bulbous, and apparently never have root tubers. Corms are of general occurrence in the Iridaceae, which also includes genera with rhizomes.

Tuberization is more a characteristic of a species than of a genus or family. A family will frequently contain genera with different types of storage organs, and there are cases of different storage organs within one genus, e.g. *Iris* species may have bulbs or rhizomes, and *Allium tuberosum* (Chinese chives) has a prominent rhizome in contrast to most *Allium* species which have bulbs. *Begonia* species may be fibrous-rooted (*B. semperflorens*), semi-tuberous (*B. socotrana*) or tuberous-rooted (*B. tuberhybrida*). Within the genus *Solanum*, only 100 or so of the 2000 known species have tubers, and only four of the 108 *Helianthus* species. Table 1.1 gives a list of families with examples of well-known genera with storage organs.

TYPES OF STORAGE ORGANS

There is confusion and uncertainty about definitions of storage organs for several reasons. Attempts to achieve rigorous definitive usage are foiled by the common horticultural use of the word 'bulb' for a range of clearly diverse storage organs, especially of ornamentals, in the same way as 'root' is used generally for the below-ground storage organs used for animal or human

Table 1.1. Families and some genera of well-known ornamental plants with storage organs of different types.

Monocotyledons	
Alliaceae	*Allium* (B, R) *Brodiaea* (C)
Alstroemeriaceae	*Alstroemeria* (R, RT)
Amaryllidaceae	*Galanthus* (B) *Narcissus* (B)
Araceae	*Caladium* (T) *Zantedeschia* (R, RT)
Cannaceae	*Canna* (R)
Hyacinthaceae	*Hyacinthus* (B) *Muscari* (B)
Iridaceae	*Iris* (B, R) *Gladiolus* (C) *Crocus* (C) *Freesia* (C)
Liliaceae	*Lilium* (B, R) *Tulipa* (B) *Scilla* (B)
Dicotyledons	
Begoniaceae	*Begonia* (ST, RT, R)
Bignoniaceae	*Incarvillea* (T)
Compositae	*Dahlia* (RT) *Liatris* (C)
Gesneriaceae	*Achimenes* (T) *Sinningia* (T)
Oxalidaceae	*Oxalis* (B, R)
Papaveraceae	*Corydalis* (T)
Primulaceae	*Cyclamen* (C)
Ranunculaceae	*Anemone* (R, S) *Ranunculus* (C)

ST = stem tuber, RT = root tuber, R = rhizome, B = bulb, C = corm.

food, irrespective of morphological considerations. Usage has changed with time; the word 'stock' is still widely used (defined by Bentham and Hooker's flora as 'a small portion of the summits of the previous year's roots as well as the base of the previous year's stems') but the glossary of Clapham, Tutin and Warburg's flora omits the word. Common usage adds confusion, lily-of-the-valley (*Convallaria*) are invariably sold as 'crowns' despite being rhizomes, whilst anemones are sold as corms, despite being tuberous rhizomes, mainly hypocotyl.

Basically there are three types of storage organ which relate to the morphological identity of the modified part, whether shoot (stem tubers, rhizomes, corms), root (root tubers) or leaves (bulbs) (Figs 1.1, 1.2 and 1.3). In all cases more than one organ is involved; bulbs include a compressed stem on which are inserted the swollen leaves, corms and rhizomes bear scale leaves, etc., but in most cases the identity of the modified organ is not in doubt. Two complications confuse, the first is that composite organs occur where boundaries between root, shoot and hypocotyl are unclear in the mature organ, and the second is the superimposition of another criterion to that of morphology – that of organ orientation. In general rhizomes grow plagiotropically, corms grow erect, whilst those that do neither remain in the generic category of stem tuber. Occasionally, one plant may have more than one type of organ, e.g. gladiolus has a main corm, but the cormels are stem

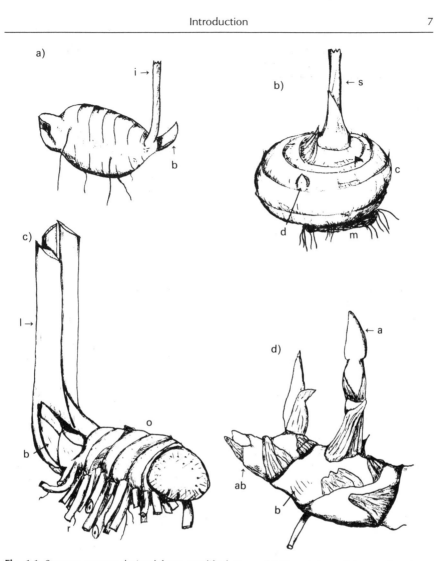

Fig. 1.1. Storage organs derived from modified stems. (a) Stem tuber of *Arum*, showing the base of the inflorescence stalk (i) and the principal vegetative bud (b) containing next year's inflorescence. (b) Corm of *Gladiolus* showing the lower portion of the aerial shoot (s), whose base is beginning to swell to form next year's corm. The remains of the old mother corm (m) is below that of the current season (c), which also has a daughter (d). (c) Pachymorph rhizome of iris. The old rhizome of the current season (o) has numerous thick roots (r), the bases of the leaves (l) enclose next year's rhizome, and the bud (b) for next year's foliage and inflorescence is lateral to the leaf bases. (d) Tip of the sympodial rhizome of *Alstroemeria* showing the aerial shoot (a) which has two scale leaves (1 and 2). The apparently terminal axillary bud (ab) grows in the axil of the first scale leaf of the previous aerial shoot (a), whose greatly enlarged basal internode (b) forms part of the rhizome, which is a chain of such internodes. Further back from the tip, roots become swollen as root tubers (not shown).

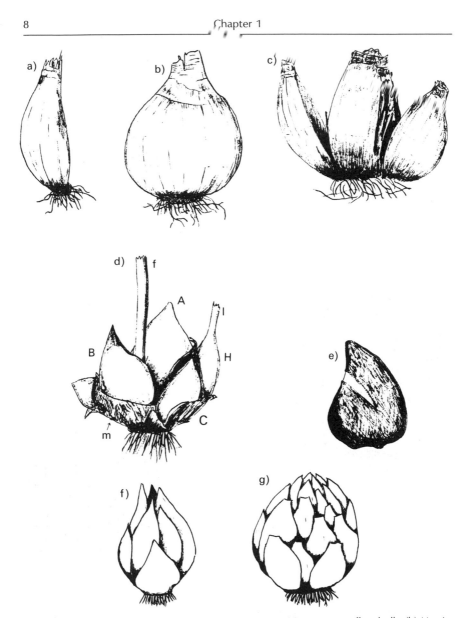

Fig. 1.2. Storage organs derived from modified leaves. (a) *Narcissus* offset bulb. (b) *Narcissus* round bulb. (c) *Narcissus* mother bulb, all three at planting time. (d) *Tulipa* cluster at lifting time showing the remains of the flower stem (f), the largest, innermost or A-bulb, followed centrifugally, and in decreasing size by the B-, C-, and D-bulbs, with externally, the H-bulb which often has its own leaf, as here (1). The mother bulb (m) has almost disappeared by this time. (e) A single tulip bulb at planting time, covered with a thin papery tunic, which is often, as here, split, to show the white scale beneath. (f) A bulb of *Lilium pumilum* with few thick, imbricate scales. (g) A bulb of *L. candidum* with numerous loose scales and leaf bases.

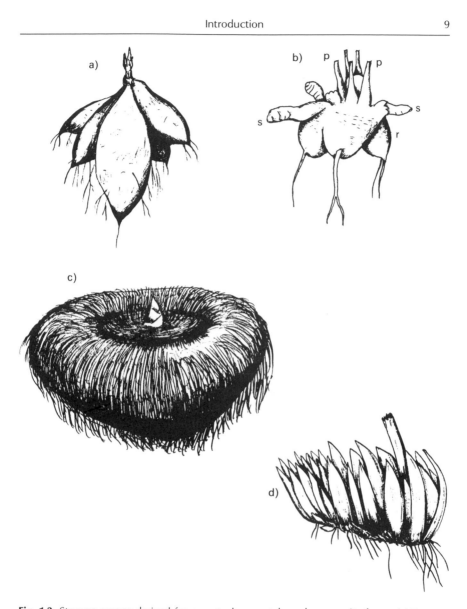

Fig. 1.3. Storage organs derived from roots, hypocotyls and composite forms. (a) Root tubers of *Dahlia*. These are joined to stem tissue at the tip, from which next year's shoots arise. (b) Young 'corm' of *Anemone* which develops from the hypocotyl and root of the seedling but becomes mainly epicotyl, comprising largely the swollen bases of the shoot with only a minor storage function by roots. The five petioles (p) surround the shoot apex, there are developing axillary shoots (s), and tuberous lateral roots (r). (c) 'Corm' of *Begonia* × *tuberhybrida* which comprises enlarged hypocotyl which persists and grows larger year by year, with additional buds forming at the centre or crown. (d) 'Bulb' of *Lilium washingtonianum* intermediate in form between an erect bulb and a rhizome.

tubers, as they lack the vertical orientation. Alstroemeria plants have rhizomes and root tubers. Intermediate forms can also cause difficulty. Some lily species have a rhizome which is plagiotropic or partly so, with swollen, non-tunicate scales, and it can be difficult to categorize an organ which is abbreviated but semi-erect.

Despite these problems, it is usually not difficult to describe storage organs in terms of the organ(s) modified for reserve storage and orientation, as follows.

Stem tuber (Fig. 1.1)

A modified stem, usually underground but identifiable as a stem by nodes and dormant buds. Aerial tubers occur in some plants, e.g. *Begonia evansiana* and some *Solanum* species, and are called *tubercles*.

Rhizome (Fig. 1.1)

A plagiotropic stem, often a main axis, either underground or on the soil surface (when it is called a *runner*, as in strawberry (where it is not a storage organ), or if very short, with a rosette appearance, an *offset*, as in *Sempervivum* (houseleek)). The term *rhizome* is useful descriptively provided its definition is not too restricted. *Stolon* is used for above- or below-ground plagiotropic stems of short duration produced by a plant with a central rosette or erect stem. A rhizome need not necessarily be modified for storing reserves, or is not so obviously so as other organs, and may serve a storage function for a single season or persist for many as in *Convallaria*. Thick fleshy rhizomes are sometimes called pachymorphic in contrast to leptomorphic, slender, ones.

Corm (Fig. 1.1)

A specialized underground stem tuber, commonly of one year's duration with a much abbreviated vertical principal axis, often formed from the base of a shoot or inflorescence, and thereby differing from the lateral stem tuber. It persists for only a single season, being replaced by another above it, as in *Gladiolus* and *Freesia*, but for several seasons in *Liatris* and in *Cyclamen*, where the corm is strictly a thickened hypocotyl. Small lateral buds produced on lateral axes are called *cormels* in *Gladiolus*, although strictly they are stem tubers. Corms are rare in the dicotyledons.

Rootstock, stock or crown

These poorly defined terms are used for structures of vertical orientation comprising either shoot or root tissues swollen in the hypocotyl region, and frequently persistent and woody or partly so. They are often used loosely as a generic name for storage or perennating parts, at or just below the soil surface. In herbaceous perennials, the crown consists of many branches, each being the base of the current year's stem. These lateral shoots grow from the base of the old stem as it dies back after flowering, and grow new adventitious roots. The new shoots flower, then die, and the process can be repeated to result in a large crown.

Root tuber (Fig. 1.3)

A swollen, modified root, often lateral, nearly always underground, as in *Dahlia*, but occasionally aerial. All root tubers have some stem tissue at the proximal end bearing buds from which regrowth occurs.

Bulb (Fig. 1.2)

A bud where leaves are modified for food storage into scales or leaf bases arranged on an abbreviated, usually vertical, axis. Bulbs can persist for a single season or several, and are usually subterranean as in *Narcissus* and *Tulipa*, aerial as in some species of *Lilium* and *Allium*, or partly buried as in other *Allium* species, or as a horticultural practice for many indoor plants in pots, e.g. *Hyacinthus*. Aerial bulbs are frequently called *bulbils*, although this word is used interchangeably for *bulblet*, which is better restricted for small bulbs. The enlarged fleshy sections of one or more nodes of orchid stems are called *pseudobulbs*.

All these categories are not recognized in other countries, e.g. French usage is restricted to *tubercule*, *rhizome* and *bulbe*, the first being subdivided into organs derived from stem, from root and from root plus hypocotyl. American usage inclines towards *tubers* for stem tubers and *tuberous roots* for root tubers, which should not lead to great confusion.

2

HORTICULTURAL USES

Despite the popular conception in the UK that all ornamental plants with bulbs and other storage organs are spring flowering garden plants, many fall outside this group, and a number of horticultural types can be readily identified as forming a somewhat artificial classification based on their usage in temperate climates – a tender plant grown in Britain as an indoor pot plant could well flourish in the open in a warmer climate. Equally, a hardy plant in Britain might not survive severe continental winters elsewhere. The geographical origin of a species is often a guide to how it can be grown, as far as any cold requirement, growing temperature and cold-hardiness are concerned. For detailed information on a wide range of genera see Mathew (1978, 1987a).

GROUPING BY FLOWERING DATE

It has been claimed that 'bulbs' are available to provide flowers in the average British garden in every month of the year, an indication of the wide range of physiological behaviour and the different plant periodicity and dormancy responses exhibited. It is also a tribute to plant collectors that such a wide range of material is available for growing in this country.

Spring Flowering Outdoor Plants

These plants include the bulbous *Narcissus, Tulipa, Hyacinthus* and *Iris*, the cormous *Crocus*, and the tuberous *Anemone*, as well as a large number of so-called 'minor' bulbs such as *Chionodoxa, Galanthus, Muscari* and *Scilla*. All are characterized by a long cold requirement, which is satisfied by the winter, the aerial parts emerging when rising temperatures in spring allow. When planted, they are all hardy in the soil of a normal UK winter, and all die down about midsummer.

Summer Flowering Plants

Of several kinds, some of these are not sufficiently hardy to withstand winter cold, and can be planted in spring only when the danger of shoots being killed by late frosts is passed, e.g. *Gladiolus* corms and *Dahlia* tubers. These are in leaf until the first hard frost of autumn. Others, such as many *Lilium* species, are winter hardy and have only a short cold requirement, initiate their flowers only after emergence of the shoots, and so flower in summer, to die down in late summer/early autumn. Summer flowering plants can be grown directly in the garden or flower bed or in pots and other containers.

Autumn Flowering Plants

A few species that flower in the autumn, often with no leaves being present (hysteranthous) include *Colchicum*, *Crocus sativus* (the saffron crocus), *Nerine* and *Sternbergia*, the first two having corms and the other two bulbs. All are hardy in the UK.

Indoor Pot Plants

There are many indoor plants that are not generally considered to be 'bulbous', as well as some that clearly are. Of the latter, the *Hippeastrum* often, but erroneously, called *Amaryllis*, is widely grown to flower in winter, as is *Cyclamen*, which has a corm. A number of aroids, such as *Caladium* and *Zantedeschia* grow from tubers, corms or rhizomes, and there are several *Begonia* species grown as pot plants which have tubers, rhizomes or corms.

USES OF 'BULBS'

Ornamentals with bulbs, corms and other storage organs are grown for a multiplicity of uses. Some provide outdoor cut flowers, others are forced for cut-flower production under glass or plastic protection, outside natural flowering dates. Many bulbs and some corms are grown for sale as pot plants, either as single species pots or as 'spring gardens' with several types in the same container. A large sector of the industry produces and sells 'dry bulbs', i.e. storage organs for sale to bulb forcers, to gardeners and for amenity horticulture use. The current interest in the environment has led many local authorities to increase the amount of 'bulb' planting in parks and public gardens, roundabouts and open spaces to produce colourful spring displays, and this is now a major outlet for 'bulb' production.

Amenity Uses

Most people know how to use 'bulbs' to good effect. This knowledge is often gained from seeing how others grow them; it is easy to be impressed by a show of flowers, the juxtaposition of species, the siting of clumps, or the complementarity of a mixed planting in a garden container, and to copy such a planting. It is necessary to have some understanding of bulbs to ensure similar success, and to maintain the effect over several seasons. A major consideration is the differing hardiness of the 'bulbs', which might necessitate planting them at different times of year, and lifting them into storage during winter; another is their different flowering dates, usually simplified into spring flowering and summer/autumn flowering. These factors, together with physiological considerations of cold requirement and preference for exposure to light, as well as the morphological ones of height, flower colour and form, provide a large matrix of choice. However, in the UK we do not have the range of eight climatic zones of the USA to make decision-making even more complicated, although we still have differences in the average severities of our winters, with more severe cold generally in the north and east of the islands. There are also considerations of shelter from wind and weather by topographic features, or, on a smaller scale, from windbreaks, hedges and walls.

Hardiness

Hardiness, or resistance to cold, is not an absolute value. It depends on acclimatization; a 'bulb' can be killed in the autumn by exposure to a low temperature which it would survive in mid-winter, after a long exposure to gradually falling temperatures. A 'bulb' can also stand lower temperatures in the dry (unplanted) condition than after planting and rehydration. Site is also important in determining the actual temperature to which the plant is exposed. Temperature at the location of a 'bulb' can be lower in a garden planter or tub than it would be in the soil nearby because of heat loss from the sides of the container. It is unusual in the UK for soil temperatures of 10 cm depth (about 'bulb' depth) to remain below 0°C for long, but the soil in a garden container, especially if off the ground on a pedestal, can freeze completely. In the Lincolnshire bulb growing district, monthly mean soil temperatures in the field at 10 cm in the coldest month of the year (which may be December, January or February) averages just over 2°C, and in none of a random 10 year run did the monthly mean fall to within a degree of freezing. Examination of the records of a 20 day unusually cold spell in February 1991 in West Sussex showed minimum screen temperatures to be below zero for all but 3 days, with a lowest temperature of −9.2°C. Grass minimum temperatures were below zero on all 20 days, the lowest temperature being − 10.9°C. At 10 cm depth, however, soil temperatures were below zero on only eight occasions, the lowest value being − 1.1°C.

There are few references to the hardiness of 'bulbs'. Dry, stored tulip bulbs have a freezing point of $c.$ − 2.5°C. After planting and acclimatization, bulbs in pots can stand − 6°C for extended periods, and − 10 to − 12°C for 24 h. Various organs have different susceptibilities to cold; roots and base plates are killed by − 13°C, but the scape and flower can withstand − 20°C. There is biochemical evidence that tulip bulbs can be regarded as 'chilling sensitive', with reduced mitochondrial respiration rates at temperatures below $c.$ 16°C.

Narcissus are hardier than iris but less so than tulip and hyacinth. A 2 day spell at − 5°C will kill unplanted narcissus bulbs, and damage is increased with progressive exposure to − 3°C, but sprouting bulbs can stand − 5 to − 10°C, depending on cultivar. However, tazetta narcissi are very much less hardy, and are not grown outdoors in the UK except in the Isles of Scilly, where frosts are unusual (averaging about three a year) and never severe. Whilst they can be grown elsewhere, especially in the Southwest, they will not survive even moderately hard winters.

Many bulbs that are planted in the autumn or remain permanently in the soil have emergent foliage or flowers during the winter. Bulbous iris, narcissi and, of course, snowdrops, are examples. These plants will survive temperatures as low as they will encounter in the UK. Flowers of narcissi will collapse at ground level at low temperatures, with the flowers lying prone on the soil surface, but regain their normal erect posture when thawed. In freak late spring cold weather, tulip flowers can fill with snow and hang down like fritillaries, but will straighten later when the snow melts.

At the other extreme, the temperature requirements of some bulbs are such that it is not worthwhile planting them in the open in this country. These must be grown in warm glasshouses or as house plants in pots, like hippeastrum, which makes them moveable enough to be transferred to the patio or outside in the summer to improve bulb growth. In between are many 'bulbs' that cannot stand winter cold, are stored in a warm location until risk of frost has passed, are then planted to flower in summer or autumn before being lifted to avoid frost damage at the approach of winter. This group includes the grandiflora gladioli and dahlias. Lists of genera in these three categories are given in Table 2.1. Further, more detailed, information can be sought in Doerflinger (1982, 1983).

Planting Situations

Outdoors, bulbs can be planted to provide good effects in many situations, especially early in the year when flowers are scarce, and a row of 'bulbs' in the vegetable garden can supply cut flowers for the house without the need to sacrifice flowers from a decorative area. In the garden, there are several

Table 2.1. List of 'bulbs' hardy, near-hardy and not hardy under UK conditions, together with suggested planting dates. Note that these are not absolute categories, and that many genera contain some species that are hardy, whilst other species are not.

	Plant in	Flower in
Hardy		
Agapanthus	spring	summer
Allium	autumn	spring, summer (almost all spp. hardy)
Alstroemeria	spring	spring/summer (*A. aurantica*)
Bloomeria	autumn	summer
Brodiaea	autumn	early summer
Bulbocodium	autumn	spring
Calochortus	autumn	early summer
Camassia	autumn	late spring
Cardiocrinum	either	summer
Chionodoxa	autumn	spring
Convallaria	late autumn	early summer
Colchicum	summer	autumn
Corydalis	late autumn	spring
Crocosmia	spring	late summer
Crocus	autumn	spring
Crocus (autumn)	summer	autumn
Cyclamen	autumn	spring, autumn (depends on species)
Eranthis	autumn	early spring
Eremurus	autumn	summer
Erythronium	autumn	spring
Fritillaria	autumn	spring
Galanthus	summer/autumn	early spring
Gladiolus (dwarf)	autumn	summer
Hemerocallis	either	summer
Hyacinthus	autumn	spring
Ipheion	autumn	spring
Iris	autumn	spring
Leucojum	spring/summer	autumn
Liatris	spring	summer
Lilium	either	summer (vast majority hardy)
Muscari	autumn	spring
Narcissus	autumn	spring (not *N. tazetta*)
Nerine	autumn	autumn (*N. bowdenii*)
Nomocharis	autumn	summer
Ornithogalum	autumn	spring/summer (*O. umbellatum*)
Oxalis (spring)	autumn	summer
Puschkinia	autumn	spring
Scilla	autumn	early spring
Sternbergia	summer	autumn
Tulipa	autumn	spring
Near-hardy (plant outside after frost danger, lift in autumn or grow in sheltered spot and mulch in winter)		
Achimenes	spring	summer
Amaryllis	summer	summer (*A. belladonna*)

Table 2.1. Cont.

	Plant in	Flower in
Anemone	either	spring/summer
Anomatheca	spring	summer
Babiana	spring	early summer
Canna	spring	summer
Chlidanthus	spring	summer
Crinum	either	summer (C. × powellii)
Curtonus	spring	summer
Dahlia	spring	summer/autumn
Eucomis	spring	summer
Freesia	spring	summer
Galtonia	spring	summer
Gladiolus	spring	summer
Habranthus	spring	summer (some species)
Hymenocallis	spring	summer
Incarvillea	spring	summer
Ixia	late autumn	summer
Ixolirion	autumn	summer
Pancratium	spring	summer
Ranunculus	autumn	spring/summer
Rhodohypoxis	spring	summer
Sauromatum	spring	spring
Schizostylis	spring	autumn
Sparaxis	either	spring/summer
Sprekelia	spring	summer
Tigridia	spring	summer
Zantedeschia	either	early summer
Zephyranthes	autumn	late summer (some species)
Not hardy[a]		
Begonia	spring	summer C, H, HB
Caladium	spring	summer C, H
Clivia	either	summer C, H
Eucharis	spring	winter
Gloriosa	spring	summer C, HB
Haemanthus	spring	summer H
Hippeastrum	autumn	winter/spring H
Lachenalia	autumn	winter H
Polianthes	spring	summer
Sandersonia	spring	summer
Sinningia	spring	summer C, H
Vallota	spring	summer C, H
Watsonia	spring	summer C, H

[a]Key to plant use: C = conservatory, H = house plant, HB = hanging basket or window box. All can be grown in a heated glasshouse, and all are grown in pots.

Fig. 2.1. Mixed summer planting of a garden border for colour, using pansies, petunias, dahlias and crocosmia (montbretia).

possibilities, including formal planting in beds, either alone or as a mixed planting with other, non-'bulb' species which are in some way complementary, either by extending the flowering duration of the plot, or by providing a contrasting or complementary colour or background at flowering time (Fig. 2.1). At the other extreme is the lack of formality in naturalized plantings of drifts of flowers in grass which are so successful on sloping land under trees (Figs 2.2 and 2.3). In between are plantings of clumps of 'bulbs' anywhere in the garden which add splashes of colour at the appropriate time in a natural way because 'bulb' plants generally grow in clumps (Fig. 2.4). A listing of broad flowering dates is given in Table 2.1, but as flowering date depends on cultivar, it will be necessary at the planning stage to seek detailed information on flowering dates from catalogues or appropriate cultivar lists.

Beds

The growing of spring 'bulbs' in beds is labour intensive because they need to be lifted after flowering to make way for summer bedding plants. To save the 'bulbs' for subsequent use, it is necessary to transfer them to another site and heel them in to allow them to die down over as long a period as possible,

Fig. 2.2. Narcissi naturalized between and under trees in a public park.

or, better still, to grow them on elsewhere for another year to allow sufficient 'bulb' growth for them to be suitable for re-use in formal planting.

Formal planting of 'bulbs' in regimented rows to give massed early colours is now much rarer in private gardens than formerly because of smaller garden sizes and the trend towards greater informality of garden design. But in public parks and gardens and associated with the gardens of national and municipal buildings the tradition continues. The selection of appropriate colours, heights and flowering dates is important for balanced effect and is also a matter of personal preference. It is also necessary to consider how well the plants will stand the exposure of the situation; tall or top-heavy flowers will not stand up to a windy site, and the display can be ruined quickly by a heavy rainstorm. Two or more cultivars of the same species can be used, such as pink and white tulips, or even a mixture of tulips of the same kind to give a wide colour range. Alternatively different genera can be used, like a subtle combination of yellow tulips and daffodils or the contrast of muscari and daffodils. When selecting 'bulbs' for a mixed planting, it is necessary that the partners should be compatible in height and flowering date – either flowering together or following on with some overlap, to avoid a flowering-gap when one partner has finished flowering and is beginning to look untidy before the other has come into bloom. The following genera are suitable for planting in beds: *Allium, Anemone, Begonia, Camassia, Canna,*

Fig. 2.3. Mixed crocuses naturalized under a large tree. Note the fence excluding the public from trampling on the plants or picking the flowers.

Chionodoxa, Colchicum, Crocus, Dahlia, Galanthus, Gladiolus, Hyacinthus, Iris, Ixia, Lilium, Muscari, Narcissus, Ornithogalum, Ranunculus, Scilla and *Tulipa*.

Plants suitable for interplanting with spring bulbs, or for use as edging, include alyssum, antirrhinum, arabis, aubretia, lobelia, myosotis, pansies, polyanthus and wallflowers. Whilst the flowering dates of the 'bulb' plants listed above and their non-bulb companions are fairly predictable, and are in about the same order in a late season as in an early one, there is an element of chance in this. In one year, for example, myosotis will flower at the same time as a given tulip cultivar, but could be 10 days later the following year, and scarcely overlap. It is a useful practical tip to set the interplants before planting the 'bulbs', to avoid damaging the latter.

Borders

Many of the considerations applying to beds are also relevant to borders, but in mixed borders with shrubs and herbaceous plants there are greater opportunities for extra contrasts, and less need for lifting the 'bulbs' (unless this is required because they are not hardy). It is of course necessary to ensure that the normal rules of border growing are applied, with taller types at the back, and the dwarf ones to the front. Many 'bulb' plants from warm climates will benefit from the extra shelter and warmth provided by a mixed border,

Fig. 2.4. Large clumps of narcissi scattered in grass with trees, in a well-known spring garden – Highdown, near Worthing.

especially if facing south, but others, native to more northerly climates or high altitudes, will require a cooler aspect to thrive. For 'bulb' plants whose flowers 'face' one way, such as narcissi and dahlias, a border against a wall or hedge, and therefore better illuminated from one direction, will ensure the maximum effect from the flowers facing outwards, towards the light.

Irregularly shaped clumps of 'bulbs' are the most effective planting arrangement; the number of 'bulbs' in a clump depends on the size of border, the size of the plants and on flower size and colour. A group of three or four orange lilies or *Fritillaria imperialis* will provide sufficient colour for all but the largest border, a dozen tulips, hyacinths, irises or narcissi will make a sufficiently large group for a medium sized border, but smaller 'bulbs' (*Crocus, Galanthus, Chionodoxa*) need to be in a group of 20–30 to make a reasonable impact. With time, some 'bulbs' (narcissus, tulip) become overcrowded and eventually fail to flower, necessitating replanting with larger, flowering sized 'bulbs' (preferably in another part of the border for disease reasons), whilst others (*Crocus, Muscari*) increase in numbers by seeding or vegetative multiplication and are best given a little extra space at planting to accommodate this behaviour. In general borders require minimum maintenance fairly frequently, and major replanting only after several years.

For public areas, planting sites for bulbs must be chosen with care to

Fig. 2.5. Narcissi in large clumps in the grass of a modern shopping area with car parking. It is essential to select planting sites carefully and to provide paths, or the emergent shoots will be inadvertently damaged before they are apparent.

avoid damage. The general public shows scant respect for grassed areas, which suffer from trampling in shopping areas; a similar fate befalls any 'bulbs' planted in the grass (Fig. 2.5). In such situations 'bulbs' survive only if in well delineated beds, or in spots where the public either cannot walk (under trees) or has no wish to do so (corners and parts with no through way).

Naturalized

Introduction of bulbs into a natural grassy situation can be most rewarding if the initial selection is of good, hardy stock. It is the simplest system of growing, requiring minimum maintenance and giving continuous flowering over many years. Orchards, banks, slopes and the edges of ponds and streams can be enhanced by large numbers of 'bulbs' growing in grass. Heavy shade is not ideal, but as many species flower and spend much of their growing season before trees are in full leaf, they can survive well in light woodland. Because the flowers appear before the grass has grown appreciably, even short 'bulbs' can grow successfully in these conditions. It is advisable to scatter the 'bulbs' on the surface before planting them, to achieve a random, natural arrangement. A bulb planter or a bulb trowel will ease the planting of larger 'bulbs', but smaller ones are most easily set by lifting a strip of turf, forking the exposed soil lightly, and replacing the turf after arranging the 'bulbs'. It is also wise to allow sufficient space for the 'bulbs' to increase naturally without becoming too overcrowded too soon. Suggested spacings are 3–6 cm apart for small 'bulbs' such as *Muscari* and *Crocus* and up to 15 cm for large ones like *Narcissus*. The usual rules about planting depth should apply; a minimum of twice the height of the 'bulb' from soil level to the base of the 'bulb'.

The following genera are recommended for naturalizing: *Allium, Anemone, Camassia, Chionodoxa, Colchicum, Crocosmia, Crocus, Fritillaria, Galanthus, Galtonia, Gladiolus, Iris, Leucojum, Lilium, Muscari, Narcissus, Ornithogalum, Scilla* and *Tulipa*. Most commonly used in the UK is *Narcissus*, and suitable mixtures of cultivars, as medium sized 'bulbs', are sold specifically for naturalizing, but many others have been grown successfully in certain situations, such as small blue or white flowered 'bulbs' (*Chionodoxa, Scilla, Muscari*) under forsythia bushes for spring colour, and snowdrops in thin woodland look so natural that it is impossible to say whether they have been planted or not.

The natural effect is emphasized by using short cultivars with simple flowers rather than heavy, showy, double-flowered forms, and avoiding bright colours which look artificial *en masse* in Britain. It is essential that the grass in which the spring 'bulbs' are growing is not mown until well after flowering, and preferably not until the foliage has completely died down. Mowing must also be stopped in early September where autumn flowering crocus and colchicum are growing. Care must also be taken not to employ standard lawn-care operations such as the use of selective herbicides, heavy rolling or lawn aeration, all of which would damage the 'bulbs'. These restrictions prevent naturalized 'bulbs' being grown where a high class lawn is required; the naturalized area must be restricted to rough grass, in orchards, or under individual trees. Naturalized 'bulbs' are also said to benefit from an annual application of fertilizer such as bone meal, especially if growing in grass where soil nutrient levels are low because of their removal in the grass mowings.

Eventually the 'bulbs' will, by natural multiplication, become overcrowded, necessitating their lifting, sorting and the replanting of the larger ones. This stage is recognized by the large amount of foliage, but declining flower numbers, which can drop alarmingly between two successive years.

An extension of naturalization is that of municipal planting of 'bulbs' in public places where the same principles are used. The enlivening of roadside verges, roundabouts and small areas of publicly owned grass with narcissi is now commonly done by local authorities using bulk purchases of bulbs to keep down costs, often supported by local businesses, local organizations such as Rotary, and involving schoolchildren in the planting. Planting such areas often costs more than the bulbs themselves. Similar plantings are successful in cemeteries and crematoria, the peripheries of car parks and similar locations where there is little danger of the plants being damaged when they emerge (Figs 2.6 and 2.7).

Fig. 2.6. Crocus in grass in a cemetery.

Fig. 2.7. Clumps of daffodils in a formal arrangement with trees in the open-plan garden of a block of flats.

Rock gardens

Many of the smaller 'bulbs' can extend the flowering season and add to the attractiveness of rock gardens, whether these are complex and large, or simple collections of a few rocks. Dwarf members of the following genera are suitable candidates for selecting rockery 'bulbs' to provide colour all year round: *Allium, Anemone, Brodiaea, Chionodoxa, Crocus, Eranthis, Erythronium, Fritillaria, Galanthus, Iris, Ixia, Muscari, Narcissus, Ornithogalum, Oxalis, Scilla, Sparaxis, Sternbergia* and *Tulipa.*

Outdoor containers

Many 'bulb' plants suitable for rock gardens are also suitable for window boxes, troughs or tubs, but an advantage of container plants is that they can be moved from some winter protection to outdoor sites as weather and season dictate. In this way, plants too delicate to spend all their lives outdoors can be grown very successfully. Container growing is also of great benefit for those with only a small garden or paved area, or even only a window sill on which to grow plants other than those inside the house. For success, attention to the growing medium is essential; garden soil alone is not as satisfactory as proprietary mixes of compost or of peat/sand mixtures because of its low fertility, and, generally, poor drainage. Depending on their siting, containers may also require frequent watering. Tubs and containers have sufficient depth to support tall plants, in contrast to window boxes which are shallow, and sometimes have to be sited in windy situations where short plants flower more successfully than tall ones. Lilies, tuberous begonias and dahlias do well in deep tubs, and can be kept out of the way until ready for transfer to a prominent position whilst in full flower. Plants in smaller pots can be buried in the garden to fill an unwanted gap, or be given a temporary site in a large container which is awaiting a more permanent plant, or summer bedding.

Frames and greenhouses

Frames, conservatories and cool greenhouses can be used to extend the growing and flowering seasons of 'bulb' plants, encourage the cultivation of more tender species, and allow flowering house plants to be brought indoors earlier. Generally, 'bulbs' in these protected conditions are grown in pots, for ease of moving them, but there is no reason not to use larger containers, for example for providing early cut flowers. The following genera can benefit from the advantage of a cool greenhouse: *Freesia, Hippeastrum, Hyacinthus, Hymenocallis, Iris, Ixia, Lilium, Muscari, Narcissus, Nerine, Oxalis, Ranunculus, Scilla, Sparaxis, Sprekelia, Tigridia, Tulipa, Vallota, Zantedeschia* and *Zephyranthes.*

For more details on selections of species for situations such as 'bulbs' for rock gardens and raised beds, for unheated frames and greenhouse, etc., consult Mathew (1987b).

PRODUCTION AREAS

Trade in 'bulbs' is international, large quantities being exported to countries where they either cannot be grown or where there is no tradition and therefore a lack of the necessary skills for growing them. Much of the trade is centred in The Netherlands, which is by far the greatest producer, but many other countries produce 'bulbs' for export as well as for internal use. Some countries specialize in individual 'bulbs'; more gladioli are grown in warmer countries, for instance, whilst the spring 'bulbs' of more temperate areas cannot be grown successfully where winters are too warm to provide the necessary low temperatures. For spring 'bulbs', the importance of fairly low temperatures (12–15°C in the growing season) and good light are recognized as favouring production; most production areas are in mild maritime climates rather than extreme continental ones. The Netherlands has a near-monopoly of world hyacinth production, and produce most of the 'minor bulbs', but the world's largest producer of narcissi is the UK. Other major producers of narcissi include the USA and Denmark. Many irises are grown in the USA, France, Israel and The Netherlands; all export some bulbs to Guernsey, a major forcer of these flowers. Tulips are mainly a European crop; areas in Japan and the UK have fallen dramatically in recent years. Tulips are increasing in popularity in Poland and in the USA, which imports most of its bulbs from The Netherlands.

It is difficult to obtain good up-to-date global information on values and areas of the various 'bulb' crops grown, which soon becomes out of date with changes in the industry. Some data on The Netherlands and the UK industries will provide an idea of their respective sizes and values. The Netherlands auction turnovers, in millions of guilders, for the major 'bulb' flowers are: tulip 225, lily 200, freesia 140 and alstroemeria 60 (£1 = c. 3.5 Dfl.). Tulip and lily are in fourth and fifth positions after rose, chrysanthemum and carnation. Alstroemeria just makes the top ten. In The Netherlands there are over 4000 'bulb' growers, producing about 8.5 thousand million 'bulbs' annually on 16 400 ha. There are 600 'bulb' exporters; with 785 million dry 'bulbs' being exported in 1989/90 plus 848 tonnes of narcissus. The largest numbers of 'bulbs' exported are, in millions, tulip 189, crocus 123, iris 100, gladioli 71, narcissus (excluding those sold by weight, above) 42, anemone 39, lily 38, hyacinth 33, freesia 25 and snowdrop 24. A further 41 genera are listed. The production value of 'bulbs' in The Netherlands is Dfl. 875 million, and the export value Dfl. 1 billion. An indication of the pre-eminence of The Netherlands in flower and plant production (not just 'bulbs'), is given by their commanding 68% of the world share of flower export (followed by

Colombia 10% and Israel 6%). For pot plants, their share is 51%, followed by Denmark (18%) and Belgium (14%).

The total UK area devoted to 'bulb' growing is *c.* 5200 ha (4400 in England and Wales), of which about 4500 are for narcissi. The value of the 'bulb' industry in England and Wales is currently about £40 million; in 1988 the breakdown was: forced 'bulb' flowers £17 million, 'bulbs' of anemone £0.2 million, of narcissi £8.6 million and of tulip £0.9 million. Values of 'bulb' flowers produced in the open were: anemone £0.9 million, iris £0.9 million, narcissi £6.1 million, and tulip £0.3 million. Total pot-plant production in England and Wales (which includes some 'bulb' flowers) was valued at £44.5 million, and that for alstroemeria £3.1 million. The total production values of all ornamentals in England and Wales for the same year were £211 million in the open and £175 million under protection. In 1990, exports from the UK (including some re-exports) totalled over 112 million 'bulbs' (103 million being narcissi), with a value of over £4 million. A third of these exports were to The Netherlands, other major destinations were Germany and the USA. However, about five times more is spent on 'bulb' imports into the UK than is received for 'bulb' exports. There is also some international trade in cut 'bulb' flowers where transport costs by air freight from warmer countries can be less than glasshouse heating costs in winter in higher latitudes. Both outdoor and forced narcissi flowers are exported from the UK to several European countries, the USA and even further afield; quantities are small and seasonally variable, but can account for up to 10% of total flower production, with a market value of several millions of pounds.

Most countries grow at least small quantities of 'bulbs', whether primary field production for cut flowers or for 'bulbs' for sale for planting in parks and gardens, or secondary flower or pot-plant production using imported 'bulbs' to supply local requirements.

3

ORIGINS, BREEDING AND SELECTION

Some 'bulb' species now grown horticulturally are almost unchanged from the wild forms, others, even of the same species, have been the subject of breeding for centuries. This has led to some highly complex modern cultivars, often based on several species, and in many cases their parentage cannot be unravelled. This is undoubtedly the reason for the within-species variability which can cause great problems for those concerned with programming flower production, necessitating accurate information to be available on the cold requirements and responses to storage treatments for each cultivar. Doorenbos (1954), in a valuable summary of the history of bulb breeding in the Netherlands, pointed out that even with the hyacinth, with a starting point of a single species of limited variability, over 5000 cultivars have been in cultivation at some time. For a large based genus, with many species, like *Narcissus*, the situation is far more complex, with lists of over 60 full species, pages of specific names of different validities, thousands of named cultivars and apparently endless duplication, a result of natural hybridization in the wild and over 300 years of deliberate crossing and selection by man. The situation is similar for the other major 'bulb' crops.

The basis for modern cultivars is the range of wild species, from which initial selections were made mainly on the basis of appearance (size of flower, shape, stem length, foliage colour and marking), and time of flowering. These selections could be made from the wild or subsequent to collection from the native habitat and introduction to northern Europe where much of the early selection was done. Deliberate crossing of plants with desirable characters was a later development.

Short accounts of some of the important genera will give an idea of their diverse origins and history.

NARCISSUS

This is a genus of the northern hemisphere, centred on Spain and Portugal, growing in a range of habitats and altitudes. It was developed as a commercial crop plant only towards the end of the 19th century. There are immense taxonomic difficulties with the genus *Narcissus* because so many species have been cultivated for a long time, there have been extensive selections from varieties and ecotypes, and there has been extensive natural hybridization in their original habitat and, more recently, deliberate crossings and selection by man. Many of these cultivars have escaped back into the wild and the true geographical origins have become confused. The Portuguese botanist Fernandes (1967) spent decades attempting to unravel relationships within the genus, based on chromosome numbers, morphology and geographical distribution. More recently, the *Flora Europaea* has condensed the number of species severely (Webb, 1980), and the Royal Horticultural Society (RHS) is revising its checklist against the Fernandes scheme, with a few modifications. The *Flora Europaea* groups the species into eight sections, as shown below, with the addition of a further section with a single species found only in Morocco, a scheme which is sufficiently logical to form an acceptable base without future major change:

1. Jonquillae with *N. fernandesii, N. gaditanus, N. jonquilla, N. requienii, N. willkommii, N. rupicola, N. cuatrecasasii, N. calcicola, N. scaberulus* and *N. viridiflorus* (plus *N. watieri* and *N. atlanticus* from Africa);
2. Narcissus with *N. poeticus* (2);
3. Serotini with *N. serotinus;*
4. Tapeinanthus with *N. humilis;*
5. Tazettae with *N. corcyrensis, N. dubius, N. elegans, N. papyraceus* (3) and *N. tazetta* (3) (plus *N. pachybolbus* from Africa);
6. Ganymedes with *N. triandrus* (3);
7. Bulbocodii with *N. bulbocodium* (2), *N. cantabricus* and *N. hedraeanthus* (plus *N. romieuxii* from Africa);
8. Pseudonarcissi with *N. asturiensis, N. bicolor, N. cyclamineus, N. lagoi, N. longispathus, N. minor, N. obvallaris* and *N. pseudonarcissus* (7); and
9. Aurelia with the single African species *N. broussonetii.*

This produces a total of 30 species for Europe plus a further five confined to North Africa. There is some doubt about the validity of four species: *N. fernandesii, N. corcyrensis, N. lagoi* and *N. obvallaris.* There are four intersectional hybrids: *N.* × *incomparabilis, N.* × *odorus, N.* × *medioluteus* and *N.* × *intermedius.* For simplicity I have not mentioned the many subspecies quoted, the numbers of which are indicated above in brackets following each species. Further details are available in Blanchard (1990).

Chromosome numbers of seven and multiples thereof (especially 14, 21, 28 and 42) are most frequent, but species in the sections Serotini and Tazettae

Table 3.1. Commercially important species of *Narcissus*, their distribution and chromosome numbers.

N. bulbocodium (2n = 14–42) Spain, Portugal, S.W. France, N. Africa
N. cyclamineus (2n = 14) Portugal
N. jonquilla (2n = 14) Spain, Portugal, N. Africa
N. poeticus (2n = 14–21) S. Europe, Spain to Greece
N. pseudonarcissus (2n = 14–42) France, Switzerland, N. Italy, N. Spain, UK
N. tazetta (2n = 20) Spain, N. Africa in a narrow band to China and Japan
N. triandrus (2n = 14) Spain, Portugal, Isles des Glenans

have ten and 11 chromosomes, based on five. It is speculated that present-day species arose from extinct ancestors with 14 chromosomes by gene mutation, hybridization, polyploidy and chromosome alteration or loss. Species cross very easily in the wild, producing spontaneous hybrids.

It is curious that although narcissus is indigenous to Europe, it became an important bulb crop later than many others. Increased interest towards the end of the 19th century was largely a result of the enthusiasm of British breeders like Peter Barr (1826–1909), known as the daffodil king because of his collection of 500 species, hybrids and cultivars, and the interest of the RHS in promoting the first Daffodil Conference in 1884. In the first half of this century many important modern cultivars were bred in Britain and in the Netherlands. Most cultivated species are derived from a few species, in particular *N. pseudonarcissus*, *N. poeticus* and *N. tazetta*, crosses between the first two giving single flowered progeny, and between the last two, multi-flowered hybrids. Commercially important species are listed alphabetically in Table 3.1.

Crosses between trumpet forms and *N. poeticus* resulted in the large-cupped narcissi which are the biggest group commercially; that of the small-cupped forms is more complex and less well understood.

Commercially important cultivars are not necessarily those most admired; growers are practical and pragmatic, they grow those cultivars which have the necessary characteristics to make them commercially worthwhile. Table 3.2 shows the current RHS classification of narcissi into divisions.

The commercial list of The Netherlands (the only country to keep annual records of the areas of individual cultivars grown) includes 330 narcissus cultivars, despite excluding any that are grown by only a single grower, but over half the area there (1639 ha) is devoted to only five: Carlton (18%), Ice Follies (9%), Dutch Master (8%), Golden Harvest (8%) and Tête-a-Tête (also written Tête-à-Tête) (8%). None of these is new, the dates of their registrations are, respectively, 1927, 1953, 1948, 1927 and 1949. In the UK about 35 cultivars are considered to be important, but six account for most of the area grown. They are the first four of the Dutch list above plus cvs Fortune and

Table 3.2. The divisions of narcissus cultivars and species, according to the Royal Horticultural Society, and the percentages of each kind grown in the Netherlands, on an area basis (total area in 1990–1991 was 1639 ha).

1.	Trumpet narcissi	26
2.	Long-cupped narcissi	41
3.	Short-cupped narcissi	2
4.	Double narcissi	12
5.	Triandrus narcissi	1
6.	Cyclamineus narcissi	11[a]
7.	Jonquilla narcissi	1
8.	Tazetta narcissi	4
9.	Poeticus narcissi	0.4
10.	Species, wild forms and hybrids	0.4
11.	Split corona narcissi	2
12.	Miscellaneous narcissi	0.5

In describing an individual flower, the number of the division is quoted followed by a colour coding (W = white, Y = yellow, G = green, P = pink, R = red and O = orange). The first letter gives the perianth colour, followed by a hyphen and the corona colour (divided into three zones eye, middle and rim, if these differ in colour). Thus cv. Actaea is described as 9W-YYR, indicating that it is a poeticus, with a white perianth and yellow corona with a red rim.
[a] Current Netherlands statistics include Tête-a-Tête with Cyclamineus, whereas the RHS classify it as a Division 12. Following the RHS classification, the Division 6 value falls to 3% and Division 12 increases proportionately.

Cheerfulness. Most cultivars are diploid, but some are triploid, the well-known but little-grown 'King Alfred' is tetraploid, and there are some hexaploids.

TULIPA

The ancestors of the garden tulip (*Tulipa gesneriana*) are presumed to be extinct, although it is known that they had long been cultivated in Turkey and Persia. Tulips are referred to in early Persian literature of the 12th century, although they were not known to Europeans until 1554 when seen by Busbequius, the Flemish Ambassador of Emperor Ferdinand I to Sulemein (also spelled Suleyman and Suleimein) the Magnificent, Sultan at Constantinople. He brought bulbs to Vienna in 1572, where they were grown by the famous herbalist Carolus Clusius. Tulips were being grown in quantity in The Netherlands before the end of the century, garden hybrids first arrived in England in 1582, and many flower shapes and colours are discernible in oil paintings of the early 1600s, which always included some 'broken' tulips, now known to be expressions of virus infection. These were highly prized and bought as speculation at extremely high prices, leading to the 'tulipomania' of 1636, which crashed dramatically in the following year.

The genus is mainly Asiatic, centred in the hilly country of Asia Minor, the southern Caucasus, Turkistan and Bukhara, petering out in northeast Asia, but persisting into China and Japan. Westwards it is found along the northern Mediterranean to Portugal, and from Morocco to Tripoli. The richest gene centre is in the northwestern spurs of the Himalayas, where there are about 50 species, out of a total of 100–150. Until the turn of the present century, pollination of tulips was left to natural means, but from then on controlled crossing programmes were initiated, particularly in The Netherlands, and use was made of *T. eichleri, T. fosteriana, T. greigii* and *T. kaufmanniana* to improve flower colour. From about 1910, the capacity for early forcing was recognized as an essential attribute of any new cultivar to attain commercial importance. The Darwin hybrid group was developed by crossing Darwin tulips with *T. fosteriana* and had a major impact on tulip growing and forcing, although their popularity is now declining. Spontaneous mutations, especially of colours, still play an important role in developing new forms of tulip.

Despite current lists containing about 5000 cultivar names, there are only about 800 generally available. The divisions are listed in Table 3.3, indicating the greatest popularity of the Triumph tulips. The top six cultivars for 1990/91 were Monte Carlo (DVT), Golden Apeldoorn (DTH), Apeldoorn (DTH), Prominence (T.T), Christmas Marvel (EVT) and Attila (T.T); between them these account for over a quarter of the total tulip area in The Netherlands (7068 ha). Cv. Apeldoorn and its several sports represent 10% of the area grown.

Wild tulip species can have chromosome numbers forming polyploid series, or occur as diploids, triploids or tetraploids or only as polyploids, but as polyploidy in tulips is mainly autopolyploidy, it has played only a minor role in the evolution of the genus. A few examinations of modern tulip cultivars indicate that about 90% are diploid, about 10% are triploid and less than 1% tetraploid. As species from central Asia have become more prominent in modern cultivars, with greater genetic imput from *T. fosteriana, T. greigii, T. ingens* and *T. kaufmanniana*, there has been an increase in the number of triploids. All the Darwin hybrids examined, bar 'Spring Song', are triploid and sterile. In 1989, the IVT, The Netherlands Plant Breeding Institute at Wageningen announced their first selections of tetraploid tulips which they hope to cross with early flowering selections of tetraploid *T. kaufmanniana* and *T. fosteriana*.

Table 3.3. The divisions of tulip cultivars and species, according to the *Classified List and International Register of Tulip Names*, with the percentage of the total area in the Netherlands (7068 ha in 1990/91) occupied by each.

EVT	1.	Single early	10
DVT	2.	Double early	9
ELT	3.	Single late	9
T.T	4.	Triumph	37
DTH	5.	Darwin hybrid	14
L.T	6.	Lily flowered	2
F.T	7.	Fringed	1
VFT	8.	Green flowered	0.3
P.T	9.	Parrot	2
DLT	10.	Double late	5
KAU	11.	Kaufmanniana	2
GRE	12.	Greigii	4
FOS	13.	Fosteriana	2
O.S	14.	Other species	2

The two and three letter codes are used to distinguish the kinds because tulip cultivars and species are usually listed alphabetically.

HYACINTHUS

Of the three recognized species of *Hyacinthus*, only one, *H. orientalis*, is commercially important. It occurs wild in the eastern Mediterranean and Turkey, and is commonly found now, naturalized, in an even larger area. It was introduced in the mid-16th century, and blue, white and purple forms were known then. Double-flowered cultivars were recorded in 1612. There are now some 50 important cultivars, the most popular colours being pink (43%), blue (24%) and white (18%) with smaller sales of red, yellow, purple and orange. Chromosome numbers vary tremendously; the diploid form has $2n = 16$, and there are triploids ($3n = 24$) and heteroploids (27, 25 and 23), but no true tetraploid. All possible chromosome numbers between 16 and 31 have been found, except 18.

The most popular cultivars grown in The Netherlands, which grows nearly all the world's hyacinths, are, in decreasing order, Pink Pearl, Anna Marie (pink), Delft Blue, Carnegie (white), White Pearl and Jan Bos (red). These six account for 69% of the total area grown.

IRIS

The irises of commercial importance are those with bulbs; the rhizomatous 'tall bearded' ones, although commonly grown in gardens are not forced or sold in large quantities for cut-flower production. The Xiphium group includes all the larger bulbous *Iris* species:

1. *I. boissieri* ($2n = 36$) Portugal and Spain;
2. *I. filifolia* ($2n = 34$) Morocco, Gibraltar, S. Spain;
3. *I. juncea* Libya, Tunisia, Algeria, Morocco, Sicily, S. Spain;
4. *I. lusitanica* Central Portugal, S.W. Spain;
5. *I. serotina* Morocco, S. Spain;
6. *I. taitii* N. Portugal;
7. *I. tingitana* ($2n = 22$) Algeria, Morocco, S. Spain;
8. *I. xiphioides* ($2n = 42$) Spanish and French Pyrenees;
9. *I. xiphium* ($2n = 34$) Algeria, Morocco, Spain, Portugal, S. France.

   I. lusitanica and *I. taitii* are considered by some as varieties of *I. xiphium*

The smaller Reticulata Irises (subgenus *Hermodactyloides*) include *I. danfordiae* ($2n = 27, 28$) and *I. reticulata* ($2n = 18, 20$), the former from the mountains of Turkey, and the latter with a wider distribution in Eastern Turkey, northern Iraq and in Iran. A number of hybrids have been produced, but this group has only limited commercial interest, occupying less than 5% of the total area of irises grown in The Netherlands.

English irises derive from *I. xiphioides*, and are so named because they were first cultivated near Bristol from bulbs imported from Spain. They are of little commercial importance because they cannot be forced, and are not responsive to temperature treatment of the bulbs. Spanish irises, which come from Spain and Portugal, are descended from *I. xiphium*. These two types were important at the turn of the century, but have become superseded by the Dutch irises, which are hybrids originating in Holland from several forms of the highly variable *I. xiphium* (some of which, like the yellow flowered *I. lusitanica*, have at times been accorded specific status) with *I. filifolia* and *I. tingitana*.

The current commercial range of cultivars is dominated by blue cultivars (over half the total area of over 700 ha grown in The Netherlands), then purple, white and yellow/white. This is helped by irises being the only blue cut flower available year-round. The most popular cultivars, in order, are Blue Magic (violet, despite its name), Ideal (blue), Prof Blaauw (dark blue), Blue Diamond, Telstar (blue) and Apollo (yellow/white), these six accounting for almost 70% of the total area. Ten years ago, the first two places were taken by Ideal (a sport of the once dominant Wedgwood) and Prof Blaauw, with Blue Magic only ninth. Purple Sensation, then third, has now fallen to 16th, indicating that cultivar popularity can change, and quite rapidly.

LILIUM

Lilies are among the oldest of cultivated flowers, unmistakably depicted on frescoes and pottery in Crete dating from 4000 years ago. There are about 90 species of lily, all in the northern hemisphere, and most are sufficiently hardy to be grown outdoors in Britain. Latitudinally they range from 63°N in Kamchatka to 11°N in southern India. Within North America they form two groups on either side of the Rocky Mountains; in the Old World two natural groupings are suggested, the East Asiatic, and the European and West Asiatic.

The groups recognized are four or seven, depending on the authority consulted, it being acknowledged that any grouping of species is unlikely to be acceptable to all. Mathew (1978) follows Woodcock and Stearn (1950) with four, thus:

1. Leucolirion (large trumpet forms);
2. Archelirion (large, wide-flowered forms);
3. Pseudolirium (upright, bowl shaped flowers coloured orange or red);
4. Martagon (pendulous flowers and recurving perianths, Turk's Cap types).

Baardse (1977) quotes the provisional classification of de Jong, with seven sections:

1. Lilium with *L. candidum, L. carniolicum, L. ponticum*;
2. Martagon with *L. hansonii, L. martagon*;
3. Pseudolirium with *L. canadense, L. pardalinum, L. superbum*;
4. Archelirion with *L. auratum, L. henryi, L. rubellum, L. speciosum*;
5. Leucolirion with *L. brownii, L. longiflorum, L. regale*;
6. Sinomartagon with *L. amabile, L. cernuum, L. davidii*;
7. Oxy· ·ala with *L. mackliniae, L. oxypetalum*.

Commercial growers are less concerned with details of botanical relationships than with broad physiological differences, allied to detailed information on the growing requirements of individual cultivars. Their two groups are the Asiatic hybrids (Asiatics) and the Oriental hybrids (Orientals). The Asiatics are derived from at least seven Asian species (*L. amabile, L. cernuum, L. concolor, L. dauricum, L. davidii, L. maculatum, L. tigrinum*) plus the European *L. bulbiferum*. These have the characteristics of early flowering (in 8–10 weeks), upright flowers, disease resistance and are readily propagated.

Commonly grown cultivars are Apeldoorn, Enchantment, Eurovision, Aristo, Prominence, Connecticut King, Sun Ray and Mont Blanc. The Orientals are derived from *L. auratum, L. speciosum* and *L. japonicum*, have flat and recurved flowers, and flower in 14–16 weeks. Cultivars include Uchida, Star Gazer and Osnat.

There has been a large increase in the area of lily growing in The Netherlands, from 600 ha in 1975 to nearly 2500 in 1989, and it now represents 15% of the total area of bulbs grown. Of the total area, 65% is taken by Asiatics, 16% by Orientals, and the remainder by longiflorum, speciosum and garden types. Because lily cultivars can be propagated rapidly, new cultivars soon replace old favourites, and lists of the most important cultivars change annually. The area of cv. Enchantment, the dominant cultivar only a few years ago, fell by over a half between 1990 and 1991, and it is now in sixth position. The current leaders are cvs Star Gazer (21%), Connecticut King (5%) and Snow Queen (3%), with a total of over 100 commercial cultivars grown in The Netherlands. In North America the Easter lily (*L. longiflorum*) is most important, using cvs Ace and Nellie White grown in the Pacific Northwest.

HIPPEASTRUM

This genus is poorly known in the wild; estimates of the number of species range from 50 to 65. All come from Central and South America, from Mexico and the West Indies south to Brazil and Argentina. With the possible exception of some more southerly montane species, they are not hardy in the UK. The genus was first introduced to Europe in the late 18th century (*Hippeastrum equestre* and *H. vittatum*), and the first hybrids were produced in 1799. There has been a long history of hybridization in The Netherlands since then, using the Brazilian species *H. aulicum*, *H. reginae*, *H. reticulatum*, *H. rutilum* and *H. vittatum*, and the Peruvian *H. psittacinum*, *H. leopoldii* and *H. padinum*. There are currently some 250 cultivars available commercially, in a number of categories which give an idea of the range of forms: trumpet-flowered, belladonna, leopoldii, miniature, double and orchid flowered. Colours available range from deep reds, pinks and oranges to more restrained peach, cream, white and two-tones. Although grown outdoors in warmer climates, they are better known as temperate region, indoor, winter flowering pot plants, frequently referred to as 'amaryllis'. In The Netherlands they are also grown for cut flowers.

GLADIOLUS

There are about 150 species, distributed across Africa and the countries bordering the Mediterranean; the majority are in South Africa. Some of the eastern Mediterranean species were introduced into northern Europe via Constantinople at least 500 years ago, and the South African species followed early in the 18th century when trade routes with India were first developed via the Cape.

The Colvillei hybrids, important for early flowering under glass, resulted from early crosses at Colville's nurseries in Chelsea in 1823. Further groups soon followed, like the Ramosus hybrids in Holland and the Gandavensis hybrids in Belgium, which led to the large-flowered or grandiflorus forms of today, first marketed in 1841. Smaller flowered types such as the Nanus hybrids were developed in the Channel Islands, and the Primulinus race traces its origin to the discovery of a yellow form of the species growing near Victoria Falls which was then crossed with the larger hybrids. However, the vast majority of the gladioli grown these days (all but 4% in The Netherlands) are the grandiflorus types.

As they are so easily hybridized, it is not surprising that well over 10 000 cultivars have been recorded, although most are no longer available. There are 115 cultivars in the current Netherlands commercial list, with the most popular colours, in terms of areas grown, being pink, followed in order by red, purple, white and yellow, with fewer orange, salmon and light red. There is less domination by a few cultivars than in other 'bulbous' species; the top three: White Friendship, Peter Pears (orange) and Hunting Song (light red) represent only 20% of the total Netherlands gladiolus growing area of 1972 ha.

FREESIA

Freesias were introduced to Europe from the Cape Province of South Africa at the turn of the century. All the modern cultivars arose from two species, the variable *Freesia refracta* which has about 17 variants of non-specific status, and *F. armstrongii*. The former is yellow, although there is a white form called *F. refracta alba*, and the latter is red/pink. Hybridization between these forms and species led to a wide range of colours being available by 1950. Since then the development of tetraploid freesias by Sparnaaij at IVT, Wageningen, has transformed the crop by increasing flower size, stem length and strength; these characteristics are now present in a large number of named cultivars of a wide range of colours. A detailed history of the commercial development of the genus is given by Goemans (1980).

CROCUS

There are about 80 species of crocus, dealt with in detail by Mathew (1982). The commercially important species include the saffron crocus *Crocus sativus* grown for the dye, flavour and medicinal saffron, and so outside the scope of our interest. The genus is grouped mainly by colour: blue, striped, white, yellow, chrysanthus, autumn flowering, and by species. Of the total Netherlands crocus area of 497 ha in 1990/91, 28% was of blue cultivars, 21% was chrysanthus and 17% was yellow. There were 59 cultivars, species, etc. in the list.

ALSTROEMERIA

Alstroemeria hybrids are a relatively new cut-flower crop, although the genus is well known as a garden plant. There are two centres for the genus in South America, one in Brazil and the other in Chile. The number of species is large; there are about 60 in total and 31 in Chile, where most occur in the Mediterranean climate area. The commercial hybrids are derived from species native to Chile: *Alstroemeria pelegrina, A. violacea* and *A. aurantica*, and were bred by two companies, Parigo in England and van Staaveren in the Netherlands in the 1960s. The named hybrids are protected by the breeders, and propagating material is sold under licence. Studies of chromosome numbers indicate that all the species examined so far are diploid, $2n = 16$. Of 25 cultivars examined, four were also $2n = 16$, 12 had $2n = 24$ ($3 \times$), six had $2n = 32$ ($4 \times$) and the remainder had either an extra or missing chromosome from the $3 \times$ or $4 \times$ situation. Autopolyploidy appears to be important in developing new commercial hybrids, together with the use of mutations. The range of cultivars changes rapidly, there are about 60 at present in a range of colours from bronze, orange, pink, red, yellow and lavender available from five major sources.

DAHLIA

Natives of Mexico, where there are 27 wild species, dahlias were introduced into Europe at the end of the 18th century, when seeds sent from Mexico produced plants at the Botanic Garden at Madrid. This led to the first description of three species of the genus, including the parent of the present cultivars, *Dahlia pinnata*. Interest in the plants soon spread to the rest of Europe. Now they are widely grown, all over the world, from seed to produce bedding plants, and from tubers. The main commercial interest lies in tuber production for retail sale, and there are 393 cultivars in the 1990 Netherlands commercial list grown on 365 ha; there is a wide range of plant height, flower size, types and colours, designated by names such as 'formal decorative', 'semi-cactus', 'pompon' and 'paeony-flowered'. Statistics are presented as numbers of tuber clusters, of which The Netherlands produce *c.* 52 million annually. The list of names changes from year to year, again reflecting the rate of propagation, the current leaders being cvs Park Princess (4% of the total), then Red Pigmy (3.3%) and Duet (2.3%), with no clear domination by a few cultivars as with some other 'bulbs'.

ANEMONE

There are a number of species, some of which are grown to some extent in gardens, such as *Anemone blanda*. The main ancestor of the cultivated anemone is *A. coronaria*, which grows in many Mediterranean countries and eastwards into Asia Minor. There are two main types in cultivation, the large, single-flowered de Caen and the St Brigid, which is semi-double, less colourful and less in demand by purchasers of cut flowers. There are named cultivars of de Caen anemones, which are of a single colour, but most stocks are of mixed colours, which may be marketed as strains, such as the St Piran strain, bred in Cornwall to provide improved winter-hardiness, combined with a wide colour range of high quality flowers.

MINOR 'BULBS'

Although minor bulbs are widely grown in gardens they have not, by definition, achieved the commercial importance of the main bulb crops, and have hitherto not received the same attention from breeders. However, there are now attempts in The Netherlands to increase sales of minor 'bulbs' by improving the market image of these species as 'special bulbs'. Many have an equally wide genetic base of several species, and many species are grown as such, rather than as hybrids. For example, *Chionodoxa* nomenclature is confused, with perhaps seven species (Mathew, 1987a) but there are only four commercially available species plus cultivars. *Muscari* has about 24 species, in four groups, and 16 commercially available species plus cultivars. Commercially available *Puschkinia* are two forms of a single species, and there are 14 species and cultivars of *Scilla*, a genus of confused nomenclature. The total area of these four cultivars in The Netherlands is 117 ha, where there are also *c.* 65 ha of alliums. These are listed as species and cultivars, the most widely grown being *Allium giganteum* (13 ha), followed by *A. moly* (12 ha) and *A. aflatunense* (11 ha).

Of the 20 or so species of *Nerine*, a South African autumn flowering genus, only four are important as cut flowers. In The Netherlands some 13 million flowers pass through the auctions annually, but in the UK most are grown in gardens, almost all *N. bowdenii*. As *N. sarniensis* is not hardy in the UK it is grown as a glasshouse crop. The other two commercial species are *N. undulata* and *N. mansellii*.

AIMS AND OBJECTIVES OF BREEDING

Basically the aim of the breeder is to produce improved plants, but the kind of improvement is very wide, from the visible attributes of flower colour, form and size, plant height, leaf colour, presentation and markings, to the less obvious, but equally important disease resistance, cold-hardiness, flowering date, 'bulb' yield, forceability, and the keeping quality of the cut flowers, as well as some less important aspects like scent. These are all general categories, within which there are many subdivisions of fine tuning which determine whether a new cultivar is successful over a long period of time, is moderately successful but soon dies out when superseded by something better, or fails completely to become accepted as a commercial cultivar. For a new line to be successful, it must possess a large number of desirable characteristics; failure in one major area means rejection, although it might still have value in a breeding programme to preserve its good qualities. An unattractive but disease resistant plant probably has more value to a breeding programme than a beautiful but susceptible one!

For ornamental plants, novelty *per se* has value, and new forms are appreciated and sought. A corollary of this is that there are fashions in what is appreciated and required, not only in the plants themselves, but in the way they are used in formal planting in gardens, as pot plants or as cut flowers. Such fashions can be influenced by advertising and promoting different types of plants in different situations.

In the early days of selection and breeding, accent was heavily on appearance, especially flower colour and form, attributes which are still of great importance, and which are easily assessed by the breeder at the stage of single plants, prior to bulking of stock – although he/she needs to be a good judge of what is required, and be prepared to test reaction to new types. More recently other characters have been more important, including ease of propagation, high growth rates and suitability for early forcing, again features that can be assessed fairly readily, although requiring a small stock of plants, several years of study, and detailed record keeping. As an example, Israeli breeders have recently produced new forms of miniature gladioli for winter flowering with a combination of desirable features, including an attractive colour pattern, slender stems with erect narrow leaves, and reaching anthesis in winter in 70–100 days, compared with 110–150 for standard grandiflorus types.

Breeding for disease resistance requires testing stocks by exposing them to several pathogens (often several strains) under controlled conditions. Recently it has been shown, by screening 200 cultivars and species, that tulip resistance to fusarium disease is based on additive gene action, and that some cultivars are efficient in transferring resistance to their progeny. It is a help to screening in this case that seedling resistance is correlated with that of the adult plants. Whilst amateur breeders can cope with plant appearance and performance, disease resistance, in particular, is a difficult area which has been tackled mainly by research stations until the involvement in recent years of some larger breeders, often on a cooperative basis with government funded research.

Floral Pigments

Despite the attraction of 'bulb' growing being almost entirely due to the appearance of the flowers, there have been few studies of the pigments responsible for the range of colours. Narcissus pigments are mainly carotenoids; β-carotene can exceptionally reach as much as 16% of the dry weight of the more deeply coloured coronas, but the more usual figure is a few percent. The common yellow and orange cultivars contain similar amounts of carotenoids but less β-carotene.

The situation in tulip is more complex. Early introductions were of yellow, white, red and purple forms which were open pollinated. Later, more controlled crossings were made, and wild species were also introduced with attractive bright red flowers which resulted in the Darwin hybrid group. For tulip, colour is determined by carotenoids, which cause the yellow colour, and flavonoids (glycosides). These are anthocyanins comprising anthocyanidins and flavonols linked with one or more glucose units: pelargonidin, cyanidin and delphinidin cause the colour range red–magenta–purple respectively. Other flavonols and glucose units are responsible for blueing effects. A detailed examination of over 500 cultivars and species by van Eijk *et al.* (1987) showed a range of pigment combinations. White cultivars contain only flavonols, sometimes with low carotenoid concentrations. Yellow ones generally have only carotenoids with the flavonols. Most of the pink and red cultivars have cyanidin and pelargonidin, but no delphinidin. Most orange cultivars have high carotenoid concentrations as well as anthocyanidin, whilst purple and violet ones have delphinidin and cyanidin as well as anthocyanidin. Cultivars with the same colour can have different pigments, and those with the same pigments can differ in colour, indicating the involvement of other chemical and physical factors.

Other Features

The success of plant breeding in the past is measured by the small number of cultivars that dominate many of the genera. These are highly successful all-round plants, although not always the most attractive. In many cases they also represent ends of lines because they are polyploid or heteroploid and sterile.

Whilst the goals of plant breeding are new all-round cultivars, it is usual that breeders concentrate on one characteristic, which can then be incorporated with others into finished varieties. Such a programme was that at Rosewarne, UK, on early outdoor flowering narcissi starting in 1963 (Fry, 1978). This was based on about 3000 cultivars for which detailed records were kept for many years. The eventual requirements were high yielding cultivars with good bulb increase, tall stems at cutting stage, with flowers clear of foliage and clear colours of the trumpet or large-cup types which would respond to pre-cooling and forcing. Initially, a worldwide collection was made of all the suitable early flowering cultivars to cross-pollinate each other, and with a very early cultivar, Rijnveld's Early Sensation, first registered in 1956. Between 1963 and 1977, 3600 successful crosses were made, producing a total of 73 000 seeds. It was necessary in many cases to store pollen of some cultivars until flowers of others were ready, and all receptor flowers were emasculated in bud to avoid accidental pollination. Bulb production of each seedling was assessed when lifted after 3 years, and flower characteristics when the plants flowered, from 4 to 6 years after sowing. Selections were made in the fifth to seventh year, and about 6% were retained initially, these being later reduced to 2% of the original number. By 1977 there were 400 clones being recorded and reselected, including a number of promising selections with potential commercial value which flowered up to 5 weeks earlier than the commercially available named cultivars. A great deal of information had been gained on the suitability of known cultivars as successful parents, and, as spin-off, a number of clones were produced with other characteristics such as late flowering and of intermediate height suitable for garden use. The first two clones of commercial cut-flower types, Tamara and Tamsyn, were registered in 1980 and granted Plant Breeders' Rights in 1981.

METHODS

Methods used for 'bulb' plants do not differ from those in general use. There is a major constraint in that most geophytes have a prolonged juvenile period, a consequence of the diversion of much of the photosynthate of the first few years into storage organs and their reserves. As the success or otherwise of a new cross depends largely on its flowering characteristics, it is a frustration to the breeder to have to wait so long for the flower, and extends enormously the time-scale of the standard procedures of hybridization, selfing, back-crossing and sib crossing, and delays the selection of the promising and discarding the unsuitable.

Tulips can take 4–6 years from germination to flowering; this long generation time, together with the low rate of natural vegetative reproduction, hinders a breeding programme and makes it uneconomic for private breeders. Collaboration with state institutions is valuable, with the release of partly completed material (breeding lines) incorporating valuable characteristics, which are then 'finished' by the private breeders and released as cultivars. Methods of early selection have proved useful short cuts. The annual increase in main bulb diameter and the offset number in the last year of the juvenile phase are correlated with bulb production by the adult plant, and there are possibilities for testing seedlings for disease resistance. It has been estimated that the use of pre-selection techniques can save up to one third of the time required using conventional methods.

For pest and disease resistance the first step is to examine existing cultivars, wild species and hybrids for resistance. Gladiolus resistance to *Fusarium oxysporum* f.sp. *gladioli* is one such, investigated in the USA. All species were found to be susceptible, but there were degrees of tolerance. A programme of hybridization, inoculation and evaluation identified several cultivars tolerant to the pathogen, which were then released. Unless further screening reveals resistant gladioli, it is unlikely that this traditional approach can improve matters further. There are, however, possibilities that gene transfer from allied genera resistant to the fungus (should these be found) could lead to resistant gladioli, and this characteristic could then be bred into present commercial cultivars whilst preserving their desirable attributes.

Embryo culture has been used particularly in lily breeding to overcome the problems of seeds with faulty endosperm which occur with interspecific crosses. Termed 'embryo rescue', the technique is a means of producing progeny from crosses which would otherwise not be successful. Another successful technique with some crosses is to insert pollen into the cavity of the style of the female parent (cut-style pollination) or to use an *in vitro* grafted-style technique, effecting pollination in cases where pollen tubes are unable to grow down the full length of the style. A further technique is to culture ovaries and ovules *in vitro*. Combining these techniques has improved the number of plantlets from a single cross as well as increasing the number of successful crosses between distantly related lilies.

Existing cultivars with a range of desirable characteristics are a useful starting point for developing new ones. Unfortunately, many such cultivars are sterile, and cannot be used in a conventional breeding programme, but quite small changes in their genetic make up can be valuable. Many 'bulb' plants have a high rate of natural mutation, and desirable phenotypic changes, such as flower colour, are easily spotted, and the plant can be isolated and propagated. The tulip sport 'Golden Apeldoorn' is now of greater commercial importance than 'Apeldoorn', thus providing a yellow version with all the benefits of the size, strength, yield and forcing ability of the original red triploid. The iris 'Ideal', a sport of the sterile 'Wedgwood', then the major cultivar grown, was so much better than the original that within a few years 'Wedgwood' almost completely disappeared as a commercial cultivar; only a tenth of a hectare is now grown in The Netherlands.

The induction of mutations is also practised on those plants where mutations are infrequent, as in lilies, as well as to increase the rate of mutations in others. Various mutagenic treatments are applied, such as X-rays, ionizing radiation or chemical treatment with nitrous oxide. Such treatments are usually given to small samples of plant material, such as lily scales, twin-scales or small bulbils, corms or tubers. Dose rates are important; it is necessary to produce changes but still allow the tissue to survive; the most favoured dose is that which allows half the material to survive. Solid mutants are those where all the cells of the plant developing following treatment are changed, a situation which is stable. In many cases chimerical structures arise, with the outer cell layer comprising changed cells overlying unchanged ones. Iris 'White Wedgwood' is chimerical, the blue flower cells are overlain by a colourless epidermis, one cell deep, which conceals the colour, so the flower looks white. Damage to the outer cell layer during flower development can lead to the appearance of blue spots or streaks, and in extreme cases produces an undesirable 'reversion' to blue flowers or parts of flowers.

Mutants can be produced using colchicine, to double the chromosome complement of the cells, and produce autotetraploids. This is easily achieved with lily bulb scales by immersing them in 0.05–0.10% colchicine solution for a few hours. Such lily autotetraploids have larger flowers than the corresponding diploids, but do not grow so well, and there is evidence that triploid lilies (derived from crosses of diploids and tetraploids) are more desirable, at least for greater vegetative growth.

NEW GENERA AND SPECIES

There are many genera of 'bulb' plants which could be grown as garden or pot plants or as sources of cut flowers. Remarkable developments have occurred over the past two decades in the breeding of new cultivars of the two well-known genera *Lilium* and *Alstroemeria*. Successful interspecific hybrids of *Cyrtanthus*, *Crinum* and *Nerine* and intergeneric crosses between *Amaryllis*, *Brunsvigia*, *Clivia*, *Crinum* and *Nerine* have recently been reported from South Africa. If more of the 1000 or so genera of 'bulb' plants were given the same attention, floriculture could be transformed. Possibilities are being investigated at the Bulb Research Centre, Lisse, The Netherlands (Koster, 1989). A start has been made with a literature survey, contacts with botanic gardens, research institutes, specialist growers and collectors and establishing a plant collection for evaluation by growers. (See also Chapter 10.)

PRESERVATION OF GENETIC MATERIAL

Efforts are made to conserve genetic material in the form of collections of cultivars of 'bulb' species of importance to horticulture. Probably the most important collection is that of *Hortus bulborum*, at Limmen in the province of Noord-Holland in The Netherlands, a private foundation where about 1100 old cultivars, some dating back to the 17th century, are carefully grown and preserved so that their special characteristics are not lost to future breeding programmes. For tulips, the collection was recently examined using a set of 20 flower, seven leaf and four stem descriptors developed as a means of reviewing and assessing the suitability of individual cultivars for specific breeding purposes.

In the UK this conservation role is the responsibility of the National

Council for the Preservation of Plants and Gardens, an independent charity based at the Royal Horticultural Society at Wisley, with links to similar organizations concerned with conservation worldwide. It has some 450 collections of genera, held by private individuals, colleges of horticulture, parks departments, nurserymen, etc. That of *Narcissus* contains 2000 taxa, and there are over 70 for *Crocus*.

There is current concern about the exploitation of wild 'bulb' plants, which are often dug up in their native habitats and are sold for export, often with their origins concealed from the purchaser who thinks they are cultivated plants grown for sale. Evidence has been found for the export of wild 'bulbs' from Afghanistan, Chile, Hungary, Iran, Italy, Japan, Nepal, Pakistan, Portugal, Saudi Arabia, Turkey and the USA (Read, 1989). The numbers of bulbs involved is large; it is claimed that in 10 years, exports from Turkey amounted to 20 million cyclamen, 71 million anemones and 111 million aconites, all largely collected from the wild.

Some species, notably *Cyclamen*, *Galanthus* and *Sternbergia*, are protected by the Convention on International Trade in Endangered Species of Wild Flora and Fauna ('CITES') now ratified by 100 countries. National legislation bans the harvesting and export of many wild flowers, but such controls are usually difficult to implement. Measures are being strengthened by improved labelling of plant consignments to indicate their origins and assist the tighter control of importation. Four of the major supermarkets in the UK have now banned the sale of wild bulbs in their stores in response to public demand. The growing of these species horticulturally, in Turkey, is being encouraged by the Flora and Fauna Preservation Society, with financial help from the World-wide Fund for Nature, as a means of setting up local industries to compete with plant collection as a source of income. The dangers from wild plant collection are the impoverishment of local floras, a serious loss of potentially valuable genetic material and the increased risk of spread of pests and diseases.

There is some interest in the use of modern techniques to help in the preservation of endangered species or maintaining potentially useful cultivars. In South Africa research on *Sandersonia* and *Gloriosa* seeks to establish methods for their micropropagation, and lily meristems have been successfully stored in liquid nitrogen, a technique that could help preserve characteristics potentially useful for the plant breeder of less successful cultivars, which might otherwise be lost.

4

MORPHOLOGY

A knowledge of the structure of 'bulb' plants is necessary for understanding how these plants grow; this is particularly so for the storage organs themselves which reveal a diversity of form, despite their common purpose of food storage. In many respects, of course, 'bulb' plants are similar to other plants which have no overt storage organs; these points of similarity will not receive the same attention as the differences.

STORAGE ORGANS

Basically, bulbs, corms and tubers are modified shoots, roots, hypocotyls or leaves, usually underground, which store food reserves. The organ(s) modified for food differ in the several types of storage organ, as already seen in Chapter 1.

Structure

Bulbs

In true bulbs, the adaptation for food storage involves leaves, either some parts or the whole organ becoming swollen by cell division to provide the necessary volume. If the former, the leaf bases persist after the attached leaf lamina has senesced and died, in some cases for several years; if the latter, the leaves are scales, which have no laminae, although they often have a membranous extension which physically protects and supports the emergent radical leaves. Some bulbs comprise scales only, some have leaf bases only, whilst others have both. It is common practice to refer to both kinds as scales. Bulb scales can be concentric, giving a tunicate bulb as in tulip, can be almost so as in iris where the two edges of each scale meet, or can be narrow with edges that meet or just overlap, as in the imbricate (scaly) bulbs of some lilies. Tunicate bulbs are protected by tunics, either single as in tulip formed by the drying of the outer scale,

46

or multiple as in iris and narcissus where old scales and leaf bases become dry, fibrous or papery. The leaves are attached to an abbreviated, usually erect stem or base plate (also called basal plate or root plate), through which the adventitious roots grow. This abbreviated stem itself sometimes has a food storing function and can grow horizontally or partly so to resemble a rhizome. Bulbs differ in longevity, some being replaced annually (e.g. tulip, where the mother bulb dies after planting to be replaced by a number of daughter bulbs, the largest of which flower the following season) whilst others persist for several years (at least four in narcissus, whose bulb is a complex branching system which persists from year to year); there is therefore scope for a number of different 'types' of bulb. Basically, however, a bulb is an abbreviated branching system with daughter bulbs of at least one other generation present, although successive generations need not correspond to successive years, as evergreen bulbs produce several generations in each year. Mature, i.e. flowering, bulb sizes correspond roughly to the size of the plant produced, ranging from about 6 cm for small bulbs of miniature iris to 30 cm and more for hippeastrum. Note that 'bulb' sizes or grades used commercially refer to circumference, not diameter.

Corms

In contrast to bulbs, corms, tubers and rhizomes are simpler structures because there is less scope for variation. Corms are generally round structures, compressed vertically, and usually less than 10 cm across, with maximum size being related to species, e.g. those of crocus are smaller than those of gladiolus. The solid stem structure is made up of several nodes and internodes. Basally they often retain the dried and shrivelled remains of the previous year's mother corm, or a large scar indicating where it was attached. Initials of the adventitious roots form at the edge of this scar. Externally, corms bear membranous, dry scales which can be removed to reveal narrow encircling scars which indicate the internodes, separating the swollen nodes, which are of differing height depending on the extent of swelling of the different parts of the corm. The growing point of a 'dormant' corm is in the depression at the centre of the upper flattened surface, where there are axillary buds, which also occur sparingly on the internodes at the equator of the corm. Flower initiation occurs a few weeks after the emergence of the flowering shoot, after planting. The base of the shoot thickens to form the new corm on top of the old one. Contractile roots grow from the base of the new corm and pull it down into the soil to prevent the corms from emerging from the soil. Branched stolon-like structures growing from the base of the new corm develop numerous small tubers at their tips; these are called cormels.

Rhizomes

These lack the concealed complexities of bulbs and corms. Rhizomes are formed by alternate vegetative and reproductive growth, and are usually symporial with the terminal bud forming an erect, often flowering, shoot while the horizontal vegetative growth is continued by lateral shoots. These laterals arise from buds, often the axillary bud in the axil of the first leaf of each erect shoot, either contemporaneously with the flowering stalk or after it dies down. The number of internodes formed by the rhizome before further production of an aerial shoot varies from a single internode as in alstroemeria to many, as in lily-of-the-valley, which has 1, 2 and 3 year old portions. In the latter, a new rhizome branch initiated in spring forms a vegetative bud by the autumn. It renews activity in the following spring and a pair of foliage leaves unfold. A flower bud forms only in the third year, and a vegetative bud is produced in the axil of the last leaf. The flower buds are short, fat and erect, are called pips or crowns, and are often sold for forced flower production under glass. Pachymorph rhizomes, as in iris, generally grow vegetatively early in the growing season and flower later in the same year. After flowering, vegetative growth continues, with a lateral branch from a flowering rhizome. This comprises, in iris, a first year vegetative portion, and in the second year a lateral vegetative branch (which develops when the flowering shoot dies, and itself flowers a year later) and a terminal flowering shoot; rhizomes live about 2 years before becoming depleted of food reserves and dying.

Stem and root tubers

Usually multinodal and bearing external buds, stem tubers are distinguishable from root tubers which bear only fibrous roots externally. Both result from the tuberization of existing organs, the former being stolons, and the latter normal roots. Both are biennial in the sense that they are formed in one season, overwinter, and then provide reserves for next year's aerial growth. Regrowth of root tubers occurs from buds on the crown; the tubers themselves have no buds, so that a detached tuber with no bud cannot regenerate a new plant.

The tuberous begonia, cyclamen and sinningia (gloxinia) have storage organs which are modified hypocotyls, which persist from year to year and continue to increase in size annually. Eventually more than one growing apex forms, allowing the 'corm' to be divided.

Specific Examples of Bulbs

The following descriptions of some true bulbs indicate the complexity of the forms which exist. The structure of other storage organs is simpler and has been dealt with above. There has been no comparative survey of bulb types; descriptions currently available are for only a few genera.

Hippeastrum

This is the simplest type; food storage occurs only in leaf bases, as there are

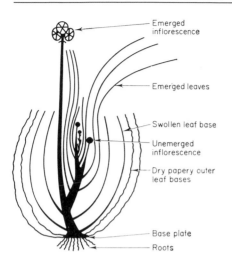

Emerged inflorescence

Emerged leaves

Swollen leaf base

Unemerged inflorescence

Dry papery outer leaf bases

Base plate

Roots

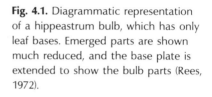

Fig. 4.1. Diagrammatic representation of a hippeastrum bulb, which has only leaf bases. Emerged parts are shown much reduced, and the base plate is extended to show the bulb parts (Rees, 1972).

no scales (Fig. 4.1). The plant is evergreen and exhibits no periodicity if grown in a heated glasshouse amply supplied with water and nutrients, except that in winter growth is slower and leaf emergence is held up (presumably a consequence of low light). The parentage of modern cultivars is complex; what follows is based on work with a single cultivar closer to wild types than the modern complex hybrids. A mature plant has 6–12 emerged leaves, and leaves are produced continuously throughout the year. Each bulb is made up of several bulb units (a sympodium or individual increment of growth, which ends in an inflorescence). Each bulb unit comprises, almost invariably, four leaves and the terminal inflorescence. Each leaf has a completely encircling base, except the innermost one which is semi-sheathing. At flower initiation, a lateral growing point forms on the side of the apex away from the last leaf, and the first leaf of this new bulb unit is on the same side as the inflorescence. The inflorescence is in the axil of the innermost, semi-sheathing leaf, and the bulb unit is in the axil of the next oldest leaf. Three bulb units are initiated each year, i.e. 12 leaves and three inflorescences. The emergence of leaves is delayed by winter, even in a heated glasshouse, but even then it is considerably more rapid than that of an inflorescence. An inflorescence elongates slowly until it is 2–3 cm long, after which it is capable of up to 6 cm a day. The time between initiation and emergence of a leaf is, depending on season, 3–8 months, that of an inflorescence 11–14 months. This difference between the growth of leaves and inflorescence is the reason why the emerged leaves present at any time are not of the same unit as the inflorescence, which therefore appears lateral to the leaf tuft. A large bulb comprises six units, two of which have lost their leaves, two bear active leaves and the leaves of the youngest two have yet to emerge. Typically it is the fourth inflorescence from the centre which is emergent. This behaviour pattern is considerably modified by the horticultural practice of drying off the bulbs in autumn and replanting them later. In particular, the third inflorescence usually dies within the bulb.

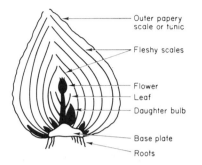

- Outer papery scale or tunic
- Fleshy scales
- Flower
- Leaf
- Daughter bulb
- Base plate
- Roots

Fig. 4.2. Diagrammatic representation of a tulip bulb at planting. It comprises scales only (i.e. the bulb has no leaf bases). Note the size order of the daughter bulbs, with the largest one innermost, and forming a decreasing size gradient centrifugally, except for the outermost one (Rees, 1972).

Lateral daughter bulbs are produced sparsely in the axils of the older, outer leaf bases, and are released as these leaf bases senesce. Such young bulbs produce only leaves, usually until nine have been formed, then an inflorescence. The first inflorescence often aborts, but the sympodial pattern is set up.

Tulipa

In contrast, the storage organs of the tulip bulb comprise scales only (Fig. 4.2). A flowering tulip plant has leaves on an aerial stem, well clear of the bulb. The base of the single leaf of a non-flowering plant forms the outer brown skin or tunic of its bulb. Tulip scales are concentric, and each has a sub-terminal hole, through which the shoot grows. A flowering size bulb has five to seven scales including the tunic. If it is a large, innermost bulb derived from a flowering mother bulb it will have a flat side corresponding to the position of the mother bulb flowering stalk. After planting, the first, lowest leaf grows on this side of the bulb. If from a non-flowering mother, a bulb is round, is called a maiden or pear, and its tunic tip indicates where the leaf was broken off. At least one daughter bulb is initiated in the axil of each scale, at no particular position, and outer scales often have more, giving an average of seven or eight per mother bulb, depending on cultivar. Not all survive; many of the small ones abort.

Narcissus

Structurally more complex is the narcissus bulb which has both scales and leaf bases as storage organs. It has similarities with hippeastrum but differs in having a more regular production of lateral bulb units, leading to a more branched system but only one set of bulb units annually. As the older parts of the bulb senesce, become papery and are sloughed, new bulb units are initiated within those formed in the previous year. The bulb unit (Fig. 4.3) comprises scales and leaves, up to five of the former and six of the latter, with the commonest combinations being 3 + 3, 4 + 3, 2 + 3 and 2 + 2, accounting for 74% of 1416 units of bulbs of cv. Fortune examined. Of the 3 + 3 combination 99% flowered, compared with 66% overall. The scales are concentric, as are the leaves, except for the innermost leaf which is

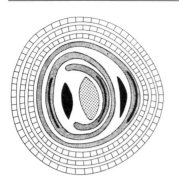

Fig. 4.3. Diagrammatic representation of a single narcissus bulb unit in transverse section. From the outside there are three scales (cross-hatched) which enclose three foliage leaves (stippled), the innermost of which has a semi-sheathing base, subtending the flower stalk (centre, dotted). In the axil of the middle leaf is next year's terminal bulb unit and in the axil of the outer leaf is next year's lateral bulb unit (both black). The positions of the leaf laminae are shown in black. (Rees, 1969. Reproduced by courtesy of *Annals of Botany*.)

semi-sheathing and next to the flower. Between the flower and the next leaf (the innermost concentric one) is a new bulb unit, and there is another bulb unit in the axil of the next leaf again (the third from the centre). The first of these is called a terminal unit (T) and the second a lateral unit (L). The former is initiated earlier, is larger and dominant. In the next season, the T unit will contain a new T unit and an L unit, but the L unit will have only a T unit.

This is the basic pattern of narcissus bulb growth, as shown in Fig. 4.4, which can be modified in a T unit by either a failure to initiate an L unit or the initiation of more than one. In the following generation, T units will contain a T and an L unit, coded TT and LT, but only in quite exceptional circumstances will an L unit contain an L unit. An examination of 252 T and L systems, and an identical number of TT and LT systems in bulbs of cv. Fortune showed that 91.7 and 90.7%, respectively were standard, 1.2 and 8.7% had no L, and 7.1 and 0.4% had an 'extra' L, remarkable consistency from a field-grown crop. Little is known about how other cultivars behave, except that some high yielding cultivars regularly seem to have extra L units. Further, whilst T units regularly produce a flower, with few exceptions, L units generally do not (Table 4.1). This regular pattern has implications for increase in bulb numbers and for increase in flower number with succeeding generations, both of which increase according to the Fibonacci series, where each number in the series is the sum of the preceding two. Thus:

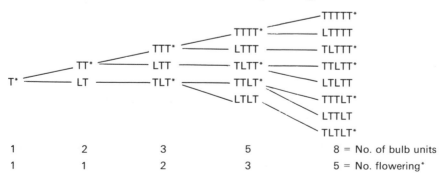

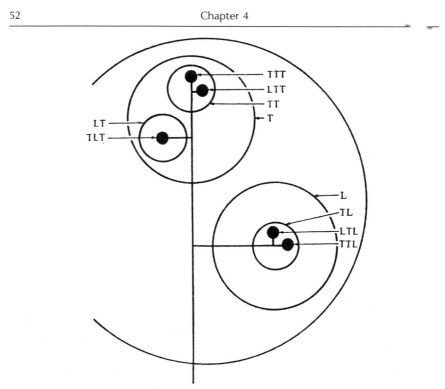

Fig. 4.4. Diagrammatic representation of a double-nosed narcissus bulb showing how the bulb can be described as a branching system of terminal (T) and lateral (L) bulb units, those of one season being shown the same diameter, and smaller than those of the previous year. In the growing season, bulb units TT, TL and LT would have leaves, but only TT and TL would have a flower. (Rees, 1969. Reproduced by courtesy of *Annals of Botany*.)

Other bulbs

From the preceding descriptions of three different bulbs, which emphasized the different organs involved, the different amounts of branching in the bulb

Table 4.1. Structure and flowering behaviour of different types of narcissus bulb unit from a single stock of cv. Fortune.

Type	No. examined	No. of scales	No. of leaves	% flowering
T	252	3.5	2.9	99
L	271	1.7	2.4	13
TT	252	3.6	2.9	98
LT	228	1.5	2.2	0
TL	272	2.1	3.5	96

Source: Rees (1969).

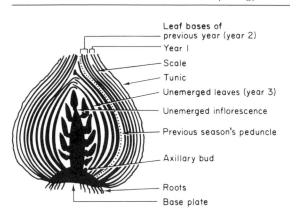

Leaf bases of
previous year (year 2)

Year I

Scale

Tunic

Unemerged leaves (year 3)

Unemerged inflorescence

Previous season's peduncle

Axillary bud

Roots

Base plate

Fig. 4.5. Diagrammatic longitudinal section of a hyacinth bulb, which is made up of leaf bases and scales (Rees, 1972).

and the longevity of the bulb units or daughter bulbs, only a brief description is necessary of some economically important bulbs. A typical mature hyacinth bulb at lifting time has the central dead inflorescence stalk surrounded by the bases of six foliage leaves and, outside these, two fleshy scales (Fig. 4.5). Surrounding these again are the leaf bases and scales of the previous years. The older leaf bases and scales are progressively thinner, more papery and pigmented purple or brown, in contrast to the younger ones which are white and swollen with storage materials. At the base of the inflorescence is the new axillary bud which will produce next season's leaves and inflorescence. Few daughter bulbs form, in contrast to the narcissus. The hyacinth bulb is almost unbranched, with almost exclusively T units.

Iris bulbs also contain scales and leaf bases. At lifting, the bulb is enclosed in a number of brown fibrous tunics, the remains of the previous season's leaves. Within are white swollen scales (usually four, the outermost being considerably larger than the others, and the innermost being a half-scale). The scales do not encircle the bulb; their edges just meet, and they are arranged oppositely. Inside is the emergent flowering shoot, enclosed by two or three sheath leaves and a number of foliage leaves. At lifting there are usually three, and this is the normal number for a non-flowering plant, which produces a central round bulb. A flowering one has up to eight or more depending on the date of planting and of flower initiation. Daughter bulbs are initiated where the scale edges meet, there being up to five in large bulbs, fewer in smaller bulbs to only one in 4–5 cm bulbs.

Lily bulbs vary in structure between species. Some, like *Lilium candidum* have radical leaves which grow in the autumn, then die, leaving bases which store reserves. Within these are true scales, which are external to the current-season radical leaves. In the axil of the innermost radical leaf alongside the emergent shoot a new daughter bulb is initiated, in late autumn. Other lilies, such as *L. martagon*, have no radical leaves; in these there are no leaf bases acting as storage organs. Flowers are initiated after shoot emergence and initiation of all the leaf primordia. During the spring the daughter bulb initiates new scales until the mother plant flowers; thereafter it grows until it is as large as the surrounding mother bulb and leaf primordia. Large mother bulbs frequently produce two daughter bulbs, and, depending on species, many lilies produce, at flowering time, small bulbils in the axils of aerial leaves as well as those underground.

The snowdrop bulb resembles that of the narcissus but is much smaller and simpler. A bulb is surrounded by only one or two outer dry brown scales, and the bulb unit comprises a single scale and two leaves, whose bases are storage organs later, as well as one or two flowers. Each scale acts as a storage organ for only 1 year. A new bulb unit (T) forms in the axil of the outer leaf, with sometimes one (L) in the same position in the leaf base of the previous year.

AERIAL PARTS

Leaves and Stems

Leaf morphology depends on the family to which the species belongs, with dicotyledonous plants generally having petiolate leaves with expanded laminae. The monocotyledonous 'bulb' plants similarly reflect family affinities, with narcissus and other amaryllids having scapes (leafless flowering stems growing from the base of the plant) and linear, radical leaves which grow by the activity of a basal meristem at the junction of the leaf blade and the basal part which remains as part of the bulb, near its tip. When the leaf dies, this meristem acts as an abscission layer. Similar leaves occur in *Muscari, Hippeastrum, Galanthus, Scilla* and *Hyacinthus*. Tulip leaves, in contrast, are borne on an erect flowering stem which may be up to 40 cm long with, usually, three or four simple, entire, lanceolate leaves, whose bases wrap around the stem. The lowest leaf is the largest, subsequent ones being progressively smaller in area. In the non-flowering tulip plant, there is only one leaf, larger than those of a flowering plant from the same sized bulb, with a petiole-like base which proximally encloses a daughter bulb, termed a maiden. Non-flowering bulbous iris plants have three leaves, and produce the commercially desired round bulbs, in contrast to the flat bulbs produced by flowering plants. Flowering plants have more leaves, usually a single radical one and up to eight on the flower stem depending on the treatment given to the bulb. The leaves are ensiform, often very narrow, distichous and folded with the silvery adaxial surfaces together. Alstroemeria leaves are curious in being inverted, by a twist in the petiole, so the abaxial surface faces upwards.

Leaves of lily can be radical or stem borne, the latter being more obvious because of their large number, up to 70 in Easter lily. Martagon lily leaves are in whorls on the lower part of the stem, but the more usual occurrence is of loosely or tightly arranged sessile, linear lanceolate leaves along the erect flowering stem. In *Lilium giganteum* the leaves are large, broad and heart shaped with long petioles. Gladiolus leaves number up to 12, are distichous and overlap at the base to give a fan-like appearance, as do those of freesia.

Flowers and Inflorescences

The flowers of some of the species of interest are described below. It is not a complete list, as this would be repetitious without adding a great deal of information, and many of the flowers are in any case sufficiently familiar not to require detailed description. A large number of geophytes have inflorescences rather than single terminal flowers, although within the same genus there may be species with both. *Narcissus tazetta* has many florets, in contrast to *N. pseudonarcissus* with a single terminal flower on an erect scape, and *Tulipa praestans* regularly has a number of flowers on a single stem, one terminal, and others on stems in the axils of the leaves. It seems likely that the single terminal flower has been derived evolutionarily from an inflorescence.

At emergence and during early growth, the narcissus scape is vertical, but near anthesis a bend forms in the pedicel just below the ovary, so the flower is held horizontally (the goose-neck stage). After fertilization, this straightens again so the fruit ripens vertically. Before the initial curvature, the flower is enclosed in a green spathe which becomes papery and brown before it is split by the final elongation of the flower. The narcissus flower comprises two whorls of perianth parts, two whorls of three introrsely dehiscing anthers in a close ring around the style, a tricarpellate inferior ovary, and the characteristic paracorolla, corona or trumpet (Fig. 4.6). The colours of the perianth parts and the corona range from white to yellow to pink and orange, and are used in classification (see Table 3.2). Nectaries between the filaments attract bumble bees, and after fertilization the ovoid green pods form. They become dry capsules by early June and dehisce into three valves to release the shiny black round seeds.

The tulip flower (Fig. 4.7) comprises six perianth parts in two equal whorls, two whorls each of three anthers and a superior tricarpellate gynoecium. There is a wide range of flower size, colour and form. The ovary has three prominent angles and three spreading stigmatic surfaces with fringed edges. The erect, cylindrical capsule splits from the top into three segments, each of which has two cavities filled with flat triangular seeds, reaching a total of up to 300 in one ovary.

Iris flowers are actinomorphic, large and initially enclosed in a spathe (Fig. 4.8). Six perianth segments are joined into a short tube above the inferior ovary. Members of the outer whorl, called falls, grow horizontally or partly erect and turn vertically downwards about a third of the distance from the tip. The inner whorl is of narrower, often erect, segments called standards. The three anthers are inserted at the bases of the falls and between them and the style branches, which are broad, petaloid, bifid at the tip and crest-like. The

Fig. 4.6. Longitudinal section through a narcissus flower. (Photograph courtesy of *Horticulture Research International*, the copyright holder.)

Fig. 4.7. Longitudinal section through a tulip flower. (Photograph courtesy of *Horticulture Research International*, the copyright holder.)

Fig. 4.8. Longitudinal section through a bulbous iris flower. (Photograph courtesy of *Horticulture Research International*, the copyright holder.)

stigmatic surface is a transverse lip near the style tip. The ovary is trilocular, and the fruit a capsule.

In the genus *Lilium*, the flowers are generally in a terminal raceme, each stalked and subtended by a bract. Each flower has six perianth segments, in two whorls, making a large, attractive bloom, frequently marked or spotted (Fig. 4.9). The form of the flower varies between species and cultivars, some being pendulous, others erect or horizontal, and there is variation in the extent of recurving of the perianth segments.

The gladiolus flower has been described as a simplified iris. The large,

Fig. 4.9. Longitudinal section through a lily flower. (Photograph courtesy of *Horticulture Research International*, the copyright holder.)

erect flowering spikes of commercial cultivars are one-sided, with all the flowers facing the same way, or arranged in two rows. Before they open each flower is enclosed by two spathes. The individual flowers are usually slightly irregular, exhibiting bilateral symmetry, with different sized perianth parts, in two whorls of three. The stigma is three-lobed, connected to the inferior ovary by a simple style. The capsule produces 50–100 seeds. There is a complete range of colours except blue, and many colour combinations, as well as of spike sizes from a few cm up to 2 m, with 30 or more flowers.

ROOTS AND ROOTING

The seedling primary root of a geophytic plant is lost during the first season's growth, being replaced by adventitious roots which persist for one season and are replaced by others from next year's storage organ. Periodic bulbs replace their root systems in late summer, the timing of root emergence and subsequent elongation apparently being controlled by soil moisture. In unplanted bulbs kept dry, root initials appear towards the end of the storage period as conical projections just under the surface of the base plate which emerge when conditions allow, and elongate rapidly. It is believe that in non-periodic bulbs, roots are produced throughout the year, maintaining a balance between leaf number and root number. In periodic bulbs the time of root production is short, and ends with shoot emergence from the bulb. Normally, no other roots are produced, so it is important that they are not damaged. Experimental excision of all roots of narcissus bulbs resulted in further root production from the base plate; when this was repeated, successively fewer roots formed, and many plants failed to flower.

Tulip root systems respond to soil type and hydrological profiles. There is usually excessive precipitation in the autumn after the bulbs are planted, and the roots are particularly sensitive to waterlogging. If damaged at this stage, flowering and the growth of daughter bulbs can be affected later in the season. For this reason well drained sand or silt type soils are usually specified for tulip growing; other 'bulb' species are less sensitive.

Few roots are produced per plant; 50–70 have been recorded in small non-flowering tulip bulbs, up to 140 in mature ones, and a mean of 32 in large hyacinth bulbs. Many bulbs and corms have roots which are unbranched or rarely so (tulip, narcissus and hyacinth) whilst others have well branched roots with second order laterals (iris, hippeastrum, lily), and root hairs are generally absent in soil- or compost-grown plants, although they can occur, sparingly, in other conditions.

Little information exists on normal rooting depths in commercial growing conditions, but tulip roots appear to be restricted to about 65 cm, whereas narcissus usually grow down to about 80 cm. Tulip roots occupy soil space more intensively than narcissus roots, probably reflecting their greater number per bulb. There are reports of mycorrhizal associations in all the bulbous species examined, including narcissus, scilla and lilies. The importance of these findings for cultivated bulbs is unknown, despite indications of better growth in lily.

Contractile roots are a feature of many geophytes, and must be important for those with corms because the daughter corm forms above the mother, tending to raise the plant out of the soil with each generation. The histological details of root contraction vary between species, but many show a sinuous contraction of the stele which results in the compression of the cortex and produces the characteristic horizontal wrinkling of the root surface. In some species, such as *Muscari* and *Bellevalia*, horizontal contractile roots move daughter bulbs away from the parent, with little change in depth relative to the soil surface (Fig. 4.10a–c). In these cases contractile roots act as dispersal organs to ensure that the daughter bulb is some distance from the mother plant rather than as a means of regulating plant depth in the soil. Some lily species, termed stoloniferous, have horizontally growing stems with bulbs at intervals along their length (Fig. 4.10d).

Other species have other means of regulating depth. In the tulip, with no contractile roots, daughter bulbs are forced deeper into the soil by 'droppers', hollow, stolon-like structures of foliar origin each of which encloses a daughter bulb near its tip (Fig. 4.10e). Extension growth occurs mainly in the zone just behind the tip and forces the daughter bulb vertically downwards. Some species and cultivars of tulip regularly produce droppers;

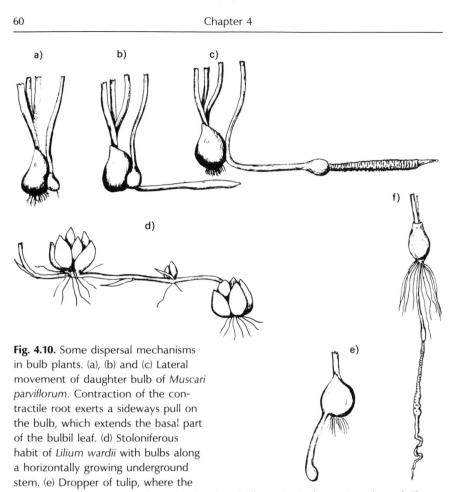

Fig. 4.10. Some dispersal mechanisms in bulb plants. (a), (b) and (c) Lateral movement of daughter bulb of *Muscari parviflorum*. Contraction of the contractile root exerts a sideways pull on the bulb, which extends the basal part of the bulbil leaf. (d) Stoloniferous habit of *Lilium wardii* with bulbs along a horizontally growing underground stem. (e) Dropper of tulip, where the elongation of the outer scale of the daughter bulb carries it deeper into the soil. The daughter bulb, oriented with the bulb tip uppermost, is within the terminal swelling of the hollow dropper. (f) *Ornithogalum nutans* whose daughter bulb dropper is pulled deeper into the soil by a contractile root. The bulb is situated in the swelling below the roots, and just above the contractile root. (All except (d) after Galil, 1961.)

in others they occur rarely, but conditions encouraging dropper formation are not well-defined. *Tulipa chrysantha* has horizontal droppers, and *Ornithogalum nutans* has a dropper-like structure associated with a contractile root (Fig. 4.10f).

Physiology

Consideration of the physiology of plants with bulbs, corms and tubers will concentrate on the special features of the group which distinguish them from other plants. Several involve the storage organs themselves: their initiation and growth, with implications for translocation and the partition of reserves and photosynthate between different plant parts, the periodicity of growth, flowering and dormancy, and the associated control mechanisms, the initiation and development of flowers and factors controlling timing of flowering. These plant processes are closely related to the environment, and are horticulturally important because manipulation of the environment (particularly temperature) both during the dormant period and during the growing phase allows accurate control of flowering, and the production of the commercial end-product.

As already seen in Chapter 4, plants with storage organs are more complex morphologically than would appear from a superficial examination. There are invariably present parts of more than one generation (up to four in narcissus), and these are at different stages of development. A treatment favouring the development of one organ might not be suitable for another in the same plant, so a great deal of perceived wisdom and standard commercial horticultural practice used in the handling of the plants has been derived empirically over many decades as a compromise between the different requirements of the various parts at any given time of year. There is also a need to distinguish specific effects of environmental factors from general ones. Temperature affects the rates of all plant processes, within limits increased temperatures speed growth and development, but there are also specific effects of temperature such as on flower initiation, on breaking 'dormancy' and on plant morphology.

'DORMANCY' AND PERIODICITY OF DEVELOPMENT

The concept of 'dormancy' is fraught with problems of definition. It can be applied broadly to indicate the absence of apparent growth, or more specifically to activities at the growing apex, which can be difficult to demonstrate, and may be caused by innate factors (i.e. be a property of the plant) or be due to external factors such as temperature or day length which have induced 'dormancy' or which are so unfavourable as to prevent growth. The terminology developed by Lang (1987) is used here; dormancy is defined as the temporary suspension of visible growth of any plant structure containing a meristem, with subdivisions of endodormancy, paradormancy and ecodormancy, where dormancy results from, respectively: i) an environmental or endogenous signal perceived by the organ itself; ii) a response to a biochemical signal from another organ; or iii) is a result of one or more unsuitable environmental factors. However, it is clear that there is considerable diversity in the morphology of bulbs and other storage organs, which often contain many meristems at different stages of activity, and that 'dormancy' may be manifest differently under natural conditions, in horticultural growing systems, or in tissue culture.

Few geophytes exhibit no 'dormancy', using the term in the horticultural sense in inverted commas, and these are typically tropical forms. However, there can be only a few areas of the world where the climate is so uniform that there are no periods unfavourable for plant growth, even if these occur only as occasional, unpredictable events. More usually, plants live in situations subject to periodic or random variations in environmental factors of different frequencies whose effects range from minor to catastrophic.

Three types of dormancy in bulbs and corms have been described by Kamerbeek et al. (1970); there may well be more. In the first, the lily and gladiolus type, the dormancy is deep, there is a cold requirement to overcome dormancy, and flowers are initiated after the start of shoot elongation and after the initiation of a fixed, high number of leaves. In the tulip type, there is hardly any true physiological dormancy, and flowers are initiated either before the aerial parts die down, as in narcissus, or soon afterwards, in the apparently 'dormant' bulb, in tulip. Cold is necessary for extension growth, and the timing of emergence and anthesis in spring is determined by ambient temperature. The third type is the bulbous iris, which has no innate, or physiological dormancy. These plants are subject to imposed summer dormancy, especially by high temperature, but grow as temperatures fall and leaves emerge in autumn. Flowers are initiated after a low temperature period and reach anthesis in spring. In the physiological sense, concerned with meristems, there is much less dormancy than might be expected from a superficial examination of plant activity, which relates 'dormancy' to lack of aerial parts.

In the horticultural sense, a bulb, corm or tuber with no emergent shoot or roots is apparently 'dormant', and therefore protected from the unfavourable environment, but examination of the apical meristem can often reveal unexpected activity; leaf primordia, roots or flowers might be in the process of being initiated, be developing or growing. Such activity illustrates the success of the survival strategy during the unfavourable period whilst simultaneously providing for the continuation of developmental processes despite the adverse conditions.

Plants with storage organs are 'dormant' (i.e. exhibit no above-ground growth) during unfavourable periods, and resume active growth when conditions allow. For many spring flowering geophytes, many of which originate in Mediterranean climates with hot, dry summers and wet, sometimes cold winters, the aerial parts die down in early summer, and the plant survives underground. The tulip is such a plant; its bulbs have a cold requirement of many weeks which must be satisfied before it can grow above ground, and this prevents emergence in the autumn although temperatures and soil moisture conditions then would be favourable for growth. Emergence in the autumn would be disastrous because the shoot would be killed by the low winter temperatures. Instead, emergence is delayed until growth is permitted as temperatures rise in spring. The plant then flowers, and, during the fairly short growing season, photosynthesis provides the food reserves for the daughter bulbs to survive the following two adverse periods of the hot, dry summer and the cold of winter.

Whilst such a pattern is common in Mediterranean climates, it is by no means universal, even within a small geographical area as shown by Schmida and Dafni (1989) for Israeli geophytes. Species have evolved different but successful strategies. Whilst the peak time for flowering of the whole of the Israeli flora is April (c. 700 species), that for geophyte flowering is a month earlier (c. 40 species). However, there is a secondary, autumn peak of flowering for geophytes, when they represent up to 80% of all the plants flowering at that time.

There are also morphological aspects of 'dormancy' and periodicity. In contrast to species in which foliage and flowers appear simultaneously (synanthous foliage), many geophytes of several families flower without leaves at the end of the summer. These hysteranthous foliage plants, including such genera as *Scilla*, *Urginea* and *Pancratium* produce leaves later in the season, often in response to the first winter rain. These plants are of two types, the *Urginea* type, and the *Crocus* type. The former have large, perennial storage organs and superterranean ovaries, the latter have smaller, annual storage organs and subterranean ovaries. The *Urginea* type characteristics are seen as being adaptations to severe and unpredictable climates, such as those bordering deserts (Dafni *et al.*, 1981).

It is clear that the possession of storage organs and their food reserves have a value as a survival mechanism, as a strategy for allowing continuing plant development, and as a means for timing processes such as flowering independently of adverse conditions. Although probably evolved from a simple system of imposed 'dormancy', whereby the factor imposing the 'dormancy' also maintains the state, simple control is unlikely to be efficient. In the near-uniform moist tropics, such a system could be satisfactory – a short dry period could induce 'dormancy', and growth would be resumed when wet conditions resumed. In a severe dry spell punctuated by a rainstorm, the renewed growth would be killed when dry conditions resumed. More complex mechanisms are more effective, either 'dormancy' breaking requiring a factor different from that inducing 'dormancy', or 'dormancy' being imposed by a factor other than the adverse one. Such factors are more effective if they have a strong seasonal element (like day length or temperature), rather than being more random, like rainfall. A simple successful 'dormancy' mechanism is that of rain flowers in the seasonal tropics which die down in the dry season but emerge, grow rapidly and flower following rain. Here there is a safety factor in that not all the flowers reach the stage of readiness to extend, and there are other flowers to respond to later rain.

Different patterns of 'dormancy' have been described for the four climatic types, Mediterranean, steppe, tropical highland and eastern Asia, which exhibit different factors affecting tuberization, different temperatures effecting 'dormancy' breaking, as well as differing seasons of sprouting and of bulbing. However, not a great deal of comparative information is available, and as distinct strategies appear to operate within a single climate, it is difficult to be certain that the categories are meaningful. The following generalizations have been made.

1. Spring flowering plants generally occur in areas of wet winters and hot, dry summers, with a short growing season during which both leaves and flowers grow above ground.
2. If winters are cold, emergence and flowering are delayed until summer.
3. Mild, wet winters followed by hot, dry summers are associated with autumn flowering.
4. If winters are cold, autumn flowering occurs, but leaf emergence is delayed until spring.

Basically it appears that leaves are above ground and photosynthesizing during the most favourable period of the year, but that flowering is less constrained to date, provided there are sufficient numbers of suitable insects available for pollination.

Horticultural advantage can be taken of a knowledge of the extent and causes of 'dormancy', the factors that release the plant from this state and the natural periodicity of a species, to devise methods of controlling its growth and flowering. Because genera behave differently, and there are also different behaviour patterns within species, it is essential that available information down to cultivar level is obtained before attempting any large-scale change in technique. Generalizations or extrapolations can be costly in terms of lost or spoiled crops.

Further, bulbs, corms and tubers are convenient items for commercial horticulture. The plant material is available in an evolutionarily designed, compact package, protected from moisture loss and provided with an ample supply of food reserves ideal for distribution and storage. In many cases the flower is already present, thus precluding the need for special treatments for its initiation.

TUBERIZATION

Because most 'bulb' plants are periodic in their development, it is not readily apparent which environmental factors are responsible for specific events such as the initiation of storage organs, and experimental modification of the environment or a change in the normal sequence of environmental factors is necessary to demonstrate a causal link.

It has been shown that the growth of daughter tulip bulbs depends on low temperatures to induce bulbing; the effect depends on the temperature, its duration and the time it is applied, relative to the stage of development of the whole plant. When bulbs with fully differentiated flower buds are cooled, lower temperatures promote rooting, shoot elongation and bulbing, the effect being more pronounced the lower the temperature, down to c. 2°C. Longer durations of the low temperature increase bulbing initially, but the process is soon saturated. If cold is applied when the shoot has one or two leaf primordia (in early June) these change into scales, and development proceeds to an apical bulb. Later transfer to cold favours bulbing rather than shoot development, and a complete range of development from apical bulbing to a normal flowering plant can be produced. These complex effects of temperature on the dormancy and development of tulip, bulbous iris and gladiolus are fully described by Le Nard (1983).

Information on the stages leading to tuberization is sparse. It appears that the modification for food storage is a process affecting an existing, 'normal' organ, a leaf, stem or root, as in the tuberization of a horizontal, unswollen stolon, sometimes in direct response to an environmental stimulus, as with the long-day response of Allium species which form a bulb. In many cases, however, it has not been fully investigated whether a newly formed organ is tuberized from its initiation, or whether this manifestation depends on the availability of food reserves being a source limitation of apparent tuberization.

Initiation and Growth of 'Bulbs'

Some reference to the position and timing of daughter 'bulbs', bulb units and other storage organs has been made in the previous chapter dealing with structure of 'bulbs'.

Corms can be 'initiated' in several ways. In gladiolus, within a few weeks of planting in spring, there is a swelling of the base of the emergent flowering shoot or on the abbreviated stem of non-flowering shoots. As the mother (planted) corm comprises several nodes, if it is sufficiently large, one or two lateral buds usually on the upper part of the corm surface also develop into new corms, and there is also a production of cormels (strictly stem tubers), often in large numbers at the ends of stolons emerging between the old and the new corms. In an investigation of the formation of daughter corms from small planted cormlets, the thickening of the internodes between the outermost sheath leaf and the first foliage leaf was followed (Yasui *et al.*, 1974). It was found that tuberization is a result of cell divisions, initially (for 9 weeks after planting) in the cortex and stele. In weeks 9–14, thickening growth is a result of cell division in the cortex and cell enlargement in both cortex and stele. Subsequently, corm enlargement is a result of cortical cell enlargement. The pattern of thickening growth in gladiolus is a form of diffuse thickening growth common in monocotyledons. When a plant is lifted at the end of summer there is one terminal replacement corm, one or more new, small lateral corms and a large number of cormels; the lateral corms and the cormels are used for propagation.

Root tubers, as in dahlia, are formed during the growing season when a root of apparently normal appearance starts to swell and become spindle shaped; tuberization in this case clearly results from the modification of an already existing organ. The rate of growth depends directly on the rate of photosynthesis by the above-ground parts, which is in turn affected by time of year, available soil moisture, etc. Several such tubers are attached to the base of the shoot, termed the crown, so they form a cluster. Tuberization is a result of cell division, initially in the cortex and pith and later in the phloem, the planes of division being such as to increase the lateral growth of the organs preferentially over growth in length, to produce the elongated swollen shape. Shoot tubers, as in ixia and the so-called cormels of gladiolus, form on the ends of subterranean stolons; like root tubers they live only one season.

Alstroemeria has a rhizome with aerial shoots in two ranks along its length. Growth is sympodial, but as the thickened apex develops a large bud, the rhizome appears to grow monopodially. The large bud is really the axillary bud in the axil of the first scale leaf of the previous aerial shoot, and the resulting chain of enlarged basal internodes forms the rhizome. A second axillary bud is also present, and can also grow, leading to a branched rhizome. It is only by the growth of this second, lateral, bud (which usually remains dormant) that there is any increase in the number of growing apices.

In true bulbs, daughter bulbs are initiated in the axils of leaves or leaf bases, in a well-defined position as in bulbous iris, where the daughter bulb occurs at the edge of each non-tunicate scale, which are inserted oppositely,

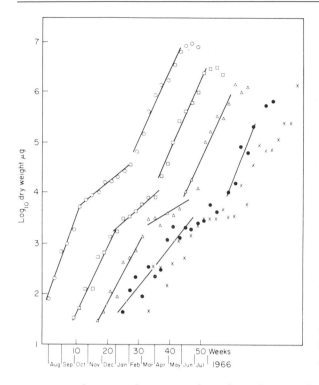

Fig. 5.1. Growth rates of tulip daughter bulbs related to their position in the mother bulb and to time of year. The time scale applies only to the innermost daughter bulb (open circles), successive daughter bulbs from the centre have each been displaced 8 weeks to the right from its inner neighbour. Note log scale. (Rees, 1968. Reproduced by courtesy of *Annals of Botany*.)

or apparently at random as in the tulip, where each scale has one or more daughter bulbs. In narcissus, with scales and leaf bases, daughter bulbs (bulb units) are initiated almost exclusively in the axils of foliage leaves. Hippeastrum daughter bulbs are initiated sparsely in the axils of outer, older leaf bases (there are no scales).

Examination of a tulip bulb at planting time in the autumn shows that this mother bulb contains daughter bulbs, and granddaughter bulbs are initiated within these, starting in the following February and continuing until July. The initiation is progressive from the axil of the outermost bulb scale inwards, and the last initiated, next to the flower, is initiated at the same time as the flower. By this time the first initiated granddaughter bulb has grown well, but after flower initiation within the daughter bulb, the innermost, small granddaughter bulbs grow most rapidly and overtake their older, outer sisters. Later on, growth of the bulbs falls into three phases, being rapid in autumn and spring, but slower in winter (Fig. 5.1). Eventually the bulb cluster at lifting contains a size series (A, B, C, . . . etc.) in which the innermost (A-bulb) is the largest, with the others progressively smaller, except for the outermost (the H-bulb, H for *huid*, Dutch for 'skin' or 'tunic') which is larger than its position would indicate. This results from the food reserves of the outermost scale being transferred to the outermost bulb late in its development as the outer scale dries to become the familiar brown skin or tunic.

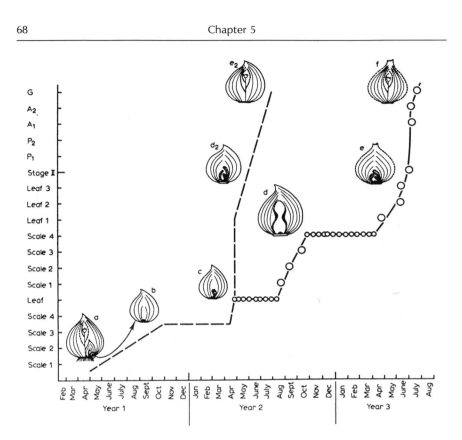

Fig. 5.2. Growth of tulip daughter bulb from initiation to flowering showing the 29 month life of the large bulb which initiates a flower in the second year and the 41 month life of one which fails to do this and lives a further 12 months. The new bulb at b is the granddaughter of the large bulb shown at a. Scale formation is complete at b, the first leaf is initiated at c, scale formation starts within the base of the single leaf at d, leaf initiation occurs at e, and flower initiation is complete at f. In the shorter lived flowering bulb, further leaf formation follows the first at d_2 and flower initiation is completed at e_2. Stage II represents the transition of the apex from vegetative to reproductive, and P, A and G are symbols for the perianth, androecium and gynoecium, respectively. (Gilford and Rees, 1974. Reproduced by courtesy of Elsevier Science Publishers BV.)

The life of a tulip bulb from its initiation until the bulb scales and flowering shoot die is about 29 months. However, if the bulb produced is too small to initiate a flower when it is in its second year, it has only a single leaf, and its life is prolonged for a further 12 months before flowering initiation to give a total life of 41 months (Fig. 5.2). During these extended periods, the daughter bulbs pass through a long spell of inactivity (ecodormancy) during each winter when there is no apical activity in producing new primordia. The 41 month bulb is also inactive in a vegetative state, with only one leaf

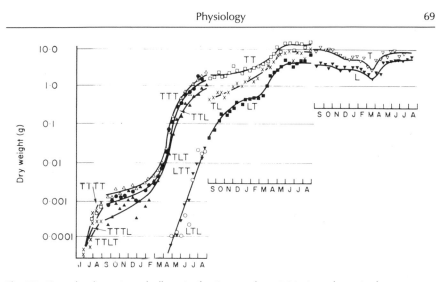

Fig. 5.3. Growth of narcissus bulb units for 3 years from initiation of terminal units, derived from one season's study of growth of different bulb units. Note the continuous growth of young bulb units even during the winter at the expense of the older ones which decrease in weight in February/March of year 3 but restore the loss in April/May, when photosynthesis resumes. Whilst there are differences between T units, they are much less than those between T and L units, the latter being initiated many months later. (Rees, 1969. Reproduced by courtesy of *Annals of Botany*.)

primordium initiated, during the summer of its second year, probably because of paradormancy.

The life span of a narcissus bulb unit is about 4 years, it being difficult to determine the end-point, but that of lateral bulb units is considerably shorter because of their later initiation (Fig. 5.3). The rate of growth of bulb units depends on season, being rapid in about March–July of each year, when the plant is photosynthesizing actively, but not during the rest of each year. However, there is evidence that young bulb units continue to grow even in September–January, at the expense of the older bulb units. Whilst the absolute growth rate of small units is low because of their small size, the older bulb units clearly lose weight to provide for growth of other parts, but regain this loss during the following active spring period. The bulb units are truly bulbous only for the latter part of their lives, after the emergence of the leaves of that bulb unit, the leaf bases and associated scales only then become swollen with food reserves. Structurally the young bulb unit resembles a spring onion before the start of bulbing. Thereafter the base of the plant, the lower ends of the leaves, which encircle the abbreviated stem and the growing point, become bulbous. Long, narrow, mature, vacuolated cells in the leaf bases become isodiametric and divide and enlarge as they become filled with storage reserves.

Implications for Yield

It has been shown for tulip that the size relationship between the bulbs making up the cluster (the product of a single planted bulb at lifting) and the total cluster weight is constant, irrespective of growing conditions and husbandry. Using this information, and the known relationship between cluster weight and the planting grade and planting density it is possible to aim for a controlled production. Thus, for cv. Apeldoorn, a 65 g cluster lifted will give a grade-out of 100 12 + cm bulbs for early forcing, 33 11–12 cm bulbs for late forcing and 97 8–11 cm bulbs for replanting for every 100 bulbs planted. To achieve 65 g clusters in an average season would require 10–11 cm bulbs planted at 25 m^{-2}. Such a model scheme could help plan growing methods with a specific end-point; ideally a yield of saleable bulbs of known grades and to provide replanting stock for continuity of the enterprise.

For narcissus, there are implications of the branching system and the weights of the different bulb units for yield prediction. The approximate weight of a mature T unit (T, TT and TL) is 30 g, that of an L unit (L and LT) is 15 g, and those of the next generation T units (TTT and TLT) are 3 g. Other combinations are too small to be of concern, and the differences between combinations like TT and TL, although real, can be ignored. The total bulb weight starting from a first year offset comprising L, TL, TTL and LTL at 48 g, becomes 81 g in the next season, followed in successive years by 129, 210, 339, 549, . . . g, again forming a Fibonacci series. But the offset bulb (48 g) of a 3 year old double-nosed (DN) bulb separates away, leaving an 81 g single-nosed (SN) bulb. The following season, the offset grows to an 81 g SN, and the SN grows to a DN, and this process continues annually. Although this is idealized, it gives an idea of pattern in the growth of narcissus bulbs, suggests potential rates of initiation and growth and could indicate shortcomings in husbandry or in crop protection measures.

Environmental Effects

In many cases there is no evidence for the involvement of environmental factors in inducing the initiation of storage organs, but standard patterns of plant growing, even outside the natural habitat, have become established which would preclude such responses being apparent. There are some well documented instances where day length is effective, as with onion, and presumably ornamental alliums, where long days (LD) are essential for bulbing, with a strong LD–temperature interaction. In alstroemeria, short days (SD) suppress rhizome formation and growth. But in most cases where the initiation or development of storage organs is affected by photoperiod, SD promote these processes, considerably enhance tuberization, reduce vegetative aerial growth and induce dormancy. Growth of both aerial and underground tubers of the tuberous-rooted (multiflora) begonias starts in SD and the whole apex eventually becomes tuberized. In dahlia, similarly, SD considerably enhance tuberization, reduce vegetative aerial growth and induce dormancy; continued development of tubers depends on continuous exposure to SD. Aerial tubers of *Begonia evansiana* are also induced by SD, although the underground tubers of the same plant have no day length response.

Investigations of possible photoperiodic effects on bulbing in narcissus and tulip have not shown any major effects. Tulip daughter bulbs exhibit a burst of growth in mid-February when there is no other obvious environmental stimulus, but no relationship between day length and initiation or growth of bulbs has been unequivocally demonstrated in narcissus, tulip, hyacinth or iris, and the formation of tulip droppers is unaffected by day length. With tulip, experiments in controlled environment cabinets showed enhanced shoot growth in LD, and the slightly, but significantly, lower bulb yield was attributed to the indirect effect of less reserve material then being available for daughter bulb growth in LD. Corm formation in *Brodiaea* is enhanced by LD.

The artificial induction of storage organs as a means of propagation will be considered in Chapter 6.

FLOWER INITIATION

Some aspects of flower initiation have already been mentioned in relation to periodicity, above, and have been reviewed (Rees, 1985). There are two main times of flower initiation relative to development; in one initiation occurs a short time after flowering, whilst the storage organ is not growing actively (in narcissus it occurs before the leaves have senesced, and in tulip in mid-summer when there are no above-ground parts, or in commercial practice when the bulbs are in storage) or is delayed until the storage organ has been replanted and the shoot has emerged (in bulbous iris, lily, gladiolus, crocus). In non-periodic evergreen plants like hippeastrum, there is continuous apical activity producing one leaf a month and an inflorescence after each fourth leaf, i.e. three a year. In one tropical species of *Pancratium* there is a new leaf each week and seven leaves between inflorescences. The timing of flower initiation seems to be related to whether floral differentiation is supported by continuous leaf activity, by the previous year's photosynthate, or is delayed until the start of next season's photosynthesis. In the third case, initiation and progress to anthesis proceed rapidly and continuously, at a rate determined by ambient temperature, culminating in winter or early spring flowering in warmer latitudes and late spring or summer flowering in colder ones.

Effect of 'Bulb' Size

An important factor determining whether or not flowering will occur is the size of the storage organ. In many corms, such as those of some aroids, crocus and gladiolus, the corm size increases annually, as each replacement corm is larger than that of the previous year (assuming good growing conditions, freedom from pests and diseases, etc.). If the corms are initially small, they do not flower, but they will eventually reach the critical size, and a flower is then initiated. In this situation flower initiation is dependent on the available food reserves; in other cases, where flowering is dependent more on current photosynthesis, and flower initiation occurs when leaves are active, this dependence is less, and the size of the planting material is less critical, as with anemone.

The situation is more complex in bulbs, because the flower is initiated in the daughter bulb, often whilst it is still growing, and dissection is necessary to establish a critical bulb unit size at the time of floral initiation. For narcissus, this was shown to be about 1 g fresh weight, with some variation between seasons, reflecting effects of growing conditions (Rees, 1986). In a mixed population of bulbs, there is a tendency to a constant flower number per tonne of bulbs for any given cultivar, although larger bulbs tend to have larger bulb units, and therefore fewer flowers per tonne, albeit of better quality. For narcissus, there is a minimum bulb weight required to ensure one flower per bulb; this is c. 10 g.

Tulips initiate their flowers later, when the daughter bulbs have completed their growth. Grading of tulip bulbs is therefore an essential part of their culture, to ensure that a given sample will, or will not, flower, as well as to guarantee flower quality, as larger bulbs produce larger flowers. The critical grade for tulip flower initiation is between 6 and 9 cm circumference, depending on cultivar. Small iris bulbs also fail to flower, the critical size being 5–8 cm, again depending on cultivar. Non-flowering tulip bulbs have only one leaf, those of iris have three. It appears that at this stage of leaf initiation the apex becomes dormant and no further primordia are initiated. When apical activity is resumed, the first primordia are scales, and a new bulb is produced. Clearly a tulip apex that goes on to produce a second leaf is committed to flowering, as is that of the iris beyond the third leaf stage. There is evidence that for iris and lily, progress beyond the critical stage towards flowering is dependent on apex size, itself directly related to bulb size.

Plant morphology plays a part in determining floral initiation, in the sense that the apex must produce a fixed or minimum number of leaves before a flower is initiated. In narcissus this number is less clearly defined than in tulip, whilst in hyacinth the change-over from the vegetative to the floral can be induced at any time by high temperature treatment.

Environmental Effects on Flowering

Temperature

This seems to be the most usual determinant of flowering, but without very critical limits beyond which flowering is prevented or cannot occur. The optimum temperature for a given species is usually quite well defined in those that have been sufficiently well studied. For tulip, the range is 5–13°C, with 17–20°C as the optimum range as indicated by earliness of initiation. Hyacinth and lily have higher optima (20–25°C) than tulip (and narcissus) and iris and allium at 9–13°C. Temperature treatment interacts with bulb size, it has long been known that with iris bulb sizes near borderline for flowering, high-temperature treatment will increase their flowering percentage, and this is used commercially for promoting flowering, where this is required.

The Easter lily has a vernalization requirement, normally provided by storing the bulbs at 2–7°C for 6 weeks. In the absence of cold, achieved by storage at 21°C from bulb harvesting, bulbs will not flower but instead produce a new bulb at the tip of the slow-growing stem after 9–12 months. Lowering the temperature before this bulb forms leads to flowering. Long days can substitute for low temperature – see later in this chapter. Roh (1989) reviewed the control of flowering in the genus *Lilium* and found that flowering in *L. longiflorum, L. speciosum* and *L. lancifolium* was accelerated by both bulb vernalization and long-day photoperiodic treatment, but at the cost of reduced flower bud number. In contrast, *L. elegans* responds only to vernalization, but had increased flower bud number.

Freesia corms have a requirement for high temperature to break their 'dormancy', termed 'aestivation'. This is normally applied to lifted, cleaned and graded corms as a 30°C heat treatment for 3 months. In the absence of such high-temperature treatment, growth is slow and irregular and flowering is delayed. Storage at 13°C results in a new corm being produced above the existing one, a process called 'pupation', which is sometimes deliberately encouraged to extend the dormant period or to give uniform planting material, as each of the new replacement corms has a single growing point. These secondary corms are also dormant and will require the extended high-temperature treatment before they will grow (Smith, 1979).

Such responses of plants which have strong periodic behaviour allow flowering to be manipulated, mainly by temperature programming, and frequently by curtailing normally extended periods of low temperature. It is essential to recognize that such treatments can be imposed only when the apical development has reached a certain morphological stage, otherwise there can be a failure of flower development, floral damage or complete floral abortion. In other cases, where there is little or no 'dormancy', provided there are no environmental restrictions to floral initiation, or these can readily be satisfied, plants can be made to grow and flower all year round. For successful flowering, bulbous iris require 9–13°C; this is a true vernalization as defined by a cold requirement for floral initiation. Bulbous iris flowers are available year-round.

There are several reported cases of morphogenic effects of temperature on 'bulb' plants, including some aberrant responses. De Munk (1989) summarized both normal and 'abnormal' effects. For hyacinth, iris and tulip, high-temperature treatment of bulbs after lifting promotes flower formation. Small bulbs, below the critical size for flowering, can be induced to flower by a high temperature or a longer exposure to high temperature. But prolonged exposure to high temperature induces leaf and flower formation in axillary buds of tulip, whilst the flower of the main shoot is killed. Low-temperature treatment of tulip allows subsequent stem elongation and rapid flowering, as described above. It also induces bulbing in vegetative buds. Hyacinth behaves similarly, although its cold requirement is less. With iris, a high temperature increases leaf number, and temperature below 20°C is required for floral initiation. De Munk concludes that these effects can be related to the stage of apical development, that of the tulip being more advanced than iris, and the transition to potential flowering has occurred in the field, whereas that of iris occurs during storage. Effects of differential day/night temperatures have been reported for *Lilium longiflorum* stem elongation, but flower and leaf lengths were related more to actual temperatures than the differential, a further example of thermomorphogenesis.

Photoperiod

In general, 'bulb' plants as a group are not greatly responsive to photoperiodic effects; the major environmental factor affecting development is temperature. Flower initiation is apparently unaffected by photoperiod in spring flowering bulb plants like narcissus, tulip, ornithogalum, iris and hyacinth or in non-periodic ones like hippeastrum. Cyclamen shows no photoperiodic response. In other species, however, there are some clear instances of effects of day length on flowering and bulbing.

In the genus *Allium*, photoperiodic responses have long been recognized, especially in onion, where LD promote bulbing and flowering. Short days promote flowering in dahlia. Most lilies have a cold requirement for flower initiation, satisfied in Easter lily by keeping moist bulbs for 6 weeks at 1.5–7°C. This cold requirement can be substituted by an LD treatment of the emerging shoot on a week-for-week basis. As the average ambient temperature in the natural habitat of the Easter lily (Lui Chiu, south of Japan, 27°N) is near 21°C, flowering is under photoperiodic control under natural conditions. A critical day length has not been defined, because the relationship between temperature and photoperiod is not well understood.

Begonia boweri, a rhizomatous species from Mexico, has been intensively investigated. The results indicate the complexity of response possible. To induce an inflorescence on the main rhizome, three or four SD are necessary, and induction is complete with three cycles of 8 h light/16 h dark, followed by 8 h light. The minimum number of SD to induce one inflorescence depends on day length; at 22°C the numbers of days of day length 11.5, 10 and 9 h are, respectively, 28–36, 8 and 4. There is also an interaction with temperature, with the upper critical day length decreasing from 12 h at 14°C to 10 h at 26°C. Continuous darkness will not induce floral initiation.

Alstroemeria plants flower earlier if subjected to LD or night break (NB) treatment, and 14 and 16 h photoperiods are more effective than 12 h ones. The total number of shoots produced decreases linearly with increase in photoperiod but the percentage of flowering shoots is greatest under a 12–16 h photoperiod. Although day lengths longer than 16 h lead to even earlier flowering, the decrease in shoot number makes it an unacceptable commercial treatment.

Many reports suggest that flower initiation in gladiolus is affected by photoperiod, but much early experimental work confounded effects of light integral, lengthened growing season and genuine photoperiodic effects. Despite early claims that gladiolus is a facultative SD plant, it now appears that flowers are initiated autonomously when vegetative growth has produced five or six leaves. LD increases flowering percentage (by reducing the number of flowers blasting), floret number per spike, and spike length, but delays flower development and anthesis. SD directly promotes corm growth, so that the photoperiodic effect can be envisaged as operating on the competition for assimilates between flower and corm. A similar effect was observed in controlled environment studies of effects of day length and NB on tulip (Hanks and Rees, 1979). Increased shoot growth in LD led to about 50% greater final stem length at anthesis; there was no effect on flowering date. The higher daughter bulb yields in SD have already been mentioned above. These results with gladiolus and tulip illustrate how photoperiodic effects, although real, can be small, are probably secondary in nature, and of little consequence for field production of 'bulbs'. The LD effect could be exploited commercially for improved flower quality in naturally short cultivars, in early forcing or in forcing under artificial lights. For pot plant production where short plants are required, SD treatment could be beneficial.

In Dutch iris, there are photoperiodic effects on flower development. Below 16°C, flower development depends on light integral rather than on photoperiod, but above this temperature, for the same light integral, more flowers abort in longer days. It is thought that the sink strength of daughter bulbs is increased by high temperatures and LD, thereby decreasing the proportion of successful flowers.

FLORAL DEVELOPMENT

For comparative purposes a system of symbols has been developed to describe the morphological stages of flower differentiation as revealed by dissection of the growing point. The scheme is sufficiently flexible to be used for all genera. With a few exceptions, development is centripetal, with perianth parts being formed before the anthers and gynoecium. For narcissus, the sequence is:

Stage I: apex flat, vegetative;
Stage II: apex dome shaped;
Stage Sp: spathe initiated;
Stage P1:   outer three perianth primordia distinguishable;
Stage P2:   inner three perianth primordia distinguishable;
Stage A1:   outer three anther primordia distinguishable;
Stage A2:   inner three anther primordia distinguishable;
Stage G: gynoecium, three carpels distinguishable;
Stage Pc: paracorolla (trumpet) apparent.

Stage Sp is not universal, but is apparent in plants which have a spathe, and Stage Pc is encountered only in genera of the Amaryllidaceae with a paracorolla, which is situated between the inner perianth whorl and the anthers. Members of the Iridaceae have only a single anther whorl, in bulbous iris this is formed before the perianth parts. Developmental stages of the most widely grown true bulbs have been illustrated by Cremer *et al.* (1974).

The time taken for the completion of flower differentiation depends on temperature and cultivar, but for tulip in the bulb growing areas of northern Europe, a typical example would be Stage I in early July and Stage G at the end of August at 17–20°C. For narcissus, where initiation begins whilst the bulbs are still in the ground, the time between Stages I and Pc is about 8 weeks, with flower differentiation being completed late in July.

The rate of floral development can be increased by temperature treatment, linked to developmental stage. It has long been known that earlier flowering of narcissus and tulip can be achieved by storing just-lifted bulbs, especially if lifted earlier than normal, at elevated temperatures. The actual temperatures and their durations appear not to be critical, and 30–35°C for 2–7 days has been used, commercially. For narcissus up to 14 days' earliness can be achieved by such warm storage (followed by 17°C) compared with storage at 17°C from lifting to the start of early cooling. A comparable benefit in tulips is 7–9 days' earliness compared with standard bulb storage at 20°C. For tulip, it appears that this benefit arises because the higher temperature stimulates leaf initiation during Stage I; this phase of development is completed sooner so that floral initiation can start earlier. From Stage G, the time to anthesis is almost constant under standard conditions, i.e. there are no seasonal effects carried over into this period of growth. Hyacinth bulbs are given high-temperature treatments after lifting and cleaning to ensure early flower initiation, and subsequently progressively lower temperatures to promote shoot and leaf elongation within the bulb. Such 'prepared' bulbs can be flowered indoors by Christmas.

Traditionally, low-temperature treatment of narcissus and tulip is applied only when flower differentiation is complete (Stage Pc in narcissus and Stage G in tulip), but fine tuning of well established procedures has indicated benefits with some tulip cultivars (including Apeldoorn) of allowing a week longer before transfer to cool conditions to minimize flower damage. For narcissus, experimental work has indicated that there is no benefit from waiting until Stage Pc, and that cooling can be started with no ill effects as early as Stage P1, allowing flowering as early as the first week of November in cv. Fortune.

A number of methods of providing low temperatures artificially to replace the natural cold of winter have been developed. They differ in degree of sophistication and of cost; these will be considered later. Suffice here to illustrate the requirements for tulip cultivars. They are not absolute; the longer the duration of low temperature the shorter will be the time to reach anthesis at a fixed high temperature. Differences between seasons are small, and the cold requirement is a characteristic of the cultivar. For 116 cultivars, the number of days at 9°C to allow the plants to flower subsequently in 21 days at 18°C ranged from 99 to 171. The requirements were in broad agreement with the International Register classification into early, mid-season and late flowering groups with a fairly wide range within each group (Rees, 1977). For three widely grown cultivars of narcissus, similar cold requirements in days at 9°C for flowering in 21 days at 16°C were within the range 123–131, but that of the early cv. Rijnveld's Early Sensation, widely used for breeding for early flowering, was only 57 (Fig. 5.4). After planting, prepared hyacinth bulbs require a low-temperature treatment of 7–10 weeks, depending on cultivar, before transfer to warm conditions to grow. They reach anthesis in 18–25 days.

There are three features of tulip growth at a high temperature (say 18°C) which are affected by failure to satisfy the cold requirement. Growth is slow, stem extension is inhibited, so the plants eventually reach anthesis on abnormally short stems (Fig. 5.5), and, within a batch, flowering dates are so variable that the normal synchronous flowering does not occur. In narcissus, stem elongation is less affected by cold treatment than in the tulip, but rate of growth and synchrony of flowering are similarly affected.

Low temperature is a requirement for successful flowering of several other species, such as *Allium karataviense*, the dwarf irises *I. danfordiae* and *I. reticulata*, *Muscari* and the crocuses. These species are generally considered along with tulips, hyacinths and narcissus because they require low-temperature growing facilities for their forcing, in contrast to other ornamentals whose storage organs can be cold-treated before planting (e.g. bulbous

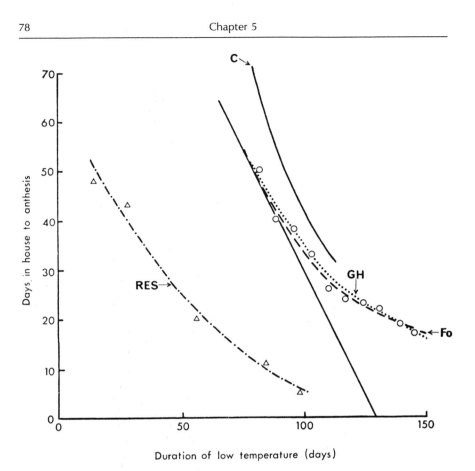

Fig. 5.4. The relationship between duration of low temperature (9°C) and the duration of the glasshouse period at 16°C for four narcissus cultivars. Three (Carlton (C), Golden Harvest (GH) and Fortune (Fo)) have quite similar responses, but that of Rijnveld's Early Sensation (RES) is markedly different and explains its earlier flowering (Rees *et al.*, 1972).

iris, lilies, lily-of-the-valley, astilbe and special tulips given the whole of the low-temperature treatment before planting). Another category of ornamental geophytes includes those which required no low-temperature treatment for successful growth and flowering, such as alstroemeria, hippeastrum, ixia, liatris, gladiolus, anemone, caladium, dahlia, ranunculus and tazetta narcissi, which includes the indoor flowering paperwhite narcissus. The culture of these plants under glass or outdoors is therefore simpler and comparable with requirements of non-tuberous species.

The environmental requirements for successful flowering of *Nerine sarniensis* was extensively studied by Warrington *et al.* (1989). These authors found that most bulbs flowered at 22°C, only a third at 14°C and all bulbs remained dormant at 30°C, until transferred to 22°C. Light intensity had less

Fig. 5.5. Effect of duration of 9°C cold (left: 6 weeks, right: 16 weeks) on subsequent stem extension at 18°C in tulip cv. Apeldoorn (Rees, 1972).

dramatic effects, but shading to reduce temperatures subsequent to flowering could result in lower bulb weight increases. This indicates that by determining and adopting optimum conditions, good flowering can be achieved even with a species known to be difficult to flower consistently.

MOBILIZATION, STORAGE AND TRANSLOCATION OF RESERVES

In comparison with food plants, little is known about this topic for ornamental plants. However, in general terms, assimilates produced by the aerial parts during the actively growing season are translocated to the storage organs until the death of the above-ground parts. These reserves are maintained during the 'dormant' period, and are remobilized when regrowth commences, i.e. the bulb acts as a sink for, then a source of, carbohydrate reserves.

The food reserves that have been studied appear to be carbohydrates of differing degrees of polymerization, synthesized from the sucrose which is the normal translocated carbohydrate. Starch is the most common stored product in the tulip bulb, accompanied by fructosyl sucrose of varying degrees of polymerization, as in other members of the Liliaceae and Amaryllidaceae; with time and temperature treatment the carbohydrate composition changes, although the significance of the changes for the development of the plant are uncertain. In general, low temperatures lead to a breakdown of complex carbohydrates into simpler molecules, including sucrose which is translocated to the shoot.

Use of $^{14}CO_2$ followed by fractionation, chromatography and autoradiography has given a broad picture of movements of carbon compounds in the tulip (Ho and Rees, 1975). Feeding leaves of tulips with $^{14}CO_2$ towards the end of the growing season (in June) indicates that sucrose is the main mobile sugar. It is translocated down the vascular bundles of the shoot to the daughter bulbs, where it accounts for the majority of the plant's radioactivity after 7 days. Once within the daughter bulb, the sucrose is progressively converted into starch and non-starch, ethanol-insoluble materials. The treated bulbs were replanted in September. Within 60 days, the growing shoots and roots were radioactive, with 10 and 14% respectively of the total plant radioactivity, and the vascular bundles strongly labelled. The labelled compounds were mainly sucrose and fructose polymers in the bulb and sucrose and glucose in the growing organs.

Further investigations of the time-course of carbohydrate movements in the plant from planting to anthesis showed the dynamic nature of the source–sink relationships in the carbon balance of the tulip (Ho and Rees, 1976, 1977). At replanting, the bulb is the sole source of carbon, with the leaves and roots being initially stronger sinks for carbon (taking 75% of the export from the scales in the first 3 months at low temperature) than daughter bulbs, flower or stem. Once the leaves are above ground and functional, they become a further source of fixed carbon, and all sink organs may receive carbon from both sources: leaves and the mother bulb. However, daughter bulbs receive a greater proportion of their supplies from the mother bulb, whilst the flower and stem receive more from the leaves. Current photosynthesis is important for tulip growth during the flowering period, and the growth of the stem and flower strongly depend on it, although this dependence is short-lived. The daughter bulbs are the strongest sink for carbon, but are a major sink only when the flower and stem have stopped growing and the scale reserves are becoming depleted. The amount of leaf assimilate supplied to the roots directly is negligible, and none goes to the old mother bulb scales. This work was done on outdoor-grown plants. Under forcing conditions, in winter under low light, development of the flower to commercial picking stage seems not to be highly dependent on current photosynthesis, and there are indications that in low light the pattern of sink strengths is changed in favour of the flower, with a suppression of daughter bulb growth.

It is likely that this pattern applies to all bulbs with the same periodicity and morphology as the tulip. Work with gladiolus agrees with the results described above and can be summarized as having two strong sinks competing for assimilates: the inflorescence and the new corm, the former being the more active until anthesis, and the latter subsequently. For bulbous plants with longer lived daughter bulbs, like narcissus, the reserves depleted from the older bulb units during the 'dormant' period and until the aerial parts are again functioning in the spring, are replenished by current photosynthesis, so the bulb units function for several years.

Where flowers are initiated after replanting in the spring, as in bulbous iris and lily, it is likely that the development of the flower, and perhaps the latter part of stem growth, is more dependent on current photosynthesis than on bulb reserves, which are mainly used for root and leaf formation and growth. If this were the case, it might explain why plants which initiate their flowers late in the growing cycle (rather than prior to or during 'dormancy') seem more dependent on light and tend to suffer from floral abortion or abscission when forced under poor winter light. For plants with corms, these storage organs are dependent on the establishment of a well developed shoot before its base becomes thickened to form the corm, so there is a priority in time for the mother corm reserves to go to roots and shoot, with a delay in daughter corm growth. In the absence of shoot growth, as in the case of untreated freesia corms, the transfer of reserves occurs to a new daughter corm, a process referred to as pupation.

PLANT GROWTH REGULATORS

Despite several examples of the involvement of plant growth regulators in the metabolism, growth and development of 'bulb' plants, there are few instances

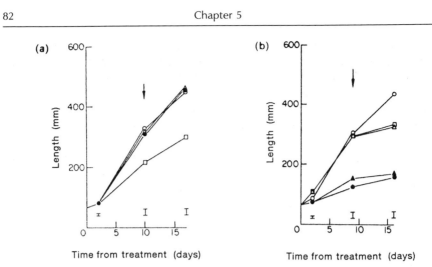

Fig. 5.6. Effects of various treatments on stem extension in tulip. (a) Excision of various parts: ○, intact controls; ●, perianth removed; △, androecium removed; □, gynoecium removed. (b) Effects of decapitation, IAA and GA: ○, intact control; ●, decapitated plus plain lanolin; △, decapitated plus IAA; ▲, decapitated plus GA; □, decapitated plus IAA and GA. Vertical bars are LSDs at $P = 0.05$ and arrows are the flowering dates of the control plants. (Hanks and Rees, 1977: Reproduced by courtesy of *New Phytologist.*)

where they have been brought into commercial practice, despite considerable research effort. Endogenous auxin, abscisic acid, cytokinins, ethylene and gibberellins (GA) have been identified in many bulbous plants.

Stem extension in both tulip and narcissus, and, by analogy, other related ornamentals, is closely related to auxin produced by the ovary (Hanks and Rees, 1977). Experiments involving the removal of different floral parts either alone or in various combinations, before the main period of stem extension, showed little effect of organs other than the ovary, whose excision in tulip was equivalent to complete flower removal in stunting extension growth, especially of the top internode. The effect of removing the ovary could be reversed almost completely by applied auxin, but not GA (Fig. 5.6). These experiments explain the restricted amount of extension growth seen in tulip plants with poor ovary development, and the very strong tall stems of the occasional plant with a quadrilocular ovary. It has been suggested that in the tulip there are two systems controlling stem extension, an auxin mediated one mainly regulating the last (top) internode and a gibberellin mediated one regulating mainly the lowest internode.

There is ample evidence that GA can replace the cold requirement of many 'dormant' seeds, buds and tubers, and it seemed an attractive idea to attempt to replace some of the long cold requirement of 'bulbs' with a quick, simple treatment which might obviate or reduce the need for cold stores, and reduce or eliminate the low-temperature period. It has long been known that the gibberellin content of tulip bulbs increases with duration of cold. Experiments have shown that whilst there is an effect of GA in some cultivars, including Apeldoorn, only part of the cold requirement can be substituted. The resultant earliness of flowering can be worthwhile, especially if combined with a shading or long-day treatment to improve stem height. With non-cooled tulips or those given very short cold treatment, GA cannot give satisfactory flowers. There is also some potential for earlier anthesis of fully cooled bulbs. The effect on height is complex; at anthesis, stems of GA treated plants are shorter than those of fully cooled controls which flower later, but final stem length is greater because GA stimulates parthenocarpic development of the gynoecium, whose auxin production results in longer top internodes (Hanks, 1982).

GA and cytokinins have been shown to reduce tulip flower losses due to floral bud blasting, dramatically in some cases. It appears that the effect is a result of an increased sink strength of the flower at a time when it is competing with the daughter bulbs for reserves from the mother bulb and for products of current photosynthesis. Although cyclamen flowers are produced early in growth (buds are formed after the sixth or seventh expanded leaf) they do not develop for several months. Elongation of their scapes and their anthesis is earlier following a GA spray to the canopy. In other species, other growth regulators have been shown to break dormancy; both cytokinins and ethylene break the dormancy of gladiolus and freesia corms. A curious effect is the formation of parrot flowers on normal tulips injected with cytokinins.

In the production of dwarfed pot plants of 'bulbs', it is usually easier to grow a vigorous cultivar and dwarf it chemically than it is to grow a naturally short one (Fig. 5.7), especially as there are effective chemicals available. For some species there are other advantages; most genetically short tulips have poor keeping quality. Ancymidol is widely used in many countries for dwarfing lilies and tulips, using low application rates such as 0.25 mg per pot of four to six tulip bulbs, but it is ineffective on narcissi and hyacinths. For these, and on grape hyacinth, ethephon can be used successfully if applied at the correct stage. If used too early there is a danger of flower blindness, and late applications are progressively less effective. A low forcing temperature impedes the dwarfing action. Effects of ancymidol on tulip can be reversed by GA, which indicates its mode of action as an anti-gibberellin. Paclobutrazol is another compound which has proved generally effective for tulip dwarfing, despite the severe flower blasting which precludes its being used on cv. Apeldoorn and its sports in some circumstances. For successful commercial culture of the tazetta narcissus cv. Grand Soleil d'Or as a pot plant, the retardants paclobutrazol and uniconazole have both been shown to be effective in reducing excessive stem and scape height. In addition to their usefulness in producing attractive pot plants of many 'bulbs', growth regulators have a role in reducing the excessive stem length of many late-forced crops, such as narcissus, so that the plants do not fall over or require mechanical

Fig. 5.7. Effects of several plant growth regulators on lily cv. Enchantment. From the left: control, cycocel at 25 000 p.p.m., phosphon at 1000 p.p.m., ethrel at 1000 p.p.m. and ancymidol at 5, 10 and 15 p.p.m. (Dicks and Rees, 1973. Reproduced by courtesy of Elsevier Science Publishers BV.)

support; stems and leaves of treated crops are more erect, stiff and short. Application can be as a compost drench or foliar spray; the former is generally preferred for narcissi and tulips, applied carefully to uniformly moist medium to avoid uneven effects on eventual plant height. For lilies, a bulb soak in a solution of growth retardant before planting is the usual method because the lily bulb's structure allows rapid and uniform uptake of the chemical.

Ethylene appears to be involved in flowering problems, especially floral blasting, bud necrosis and abscission of forced iris, lily and tulip. Ethylene can be generated in stores by tulip bulbs infected with *Fusarium*, or is released by ripening fruit or cut flowers, or from heating equipment or burners generating CO_2 for greenhouse enrichment. Ethylene also occurs naturally in the soil of waterlogged fields. Damage can result from ethylene concentrations as low as 0.1 p.p.m., and as a single *Fusarium* infected bulb can produce over $0.1 \, \text{cm}^3$ of the gas daily, it is clearly essential to have good ventilation of stores. Other deleterious effects include impeded flower and shoot development in stored tulip bulbs, the formation of gums by corms and bulbs of several species, stunted growth and inhibited rooting.

There appear to be several periods when the plant is susceptible to effects of ethylene. One is at lifting time when the shoot within the bulb is small, and another is at pollen tetrad formation, about 4 weeks after the completion of flower differentiation in the stored tulip, but after planting and when the shoot is well grown in lily. Flower bud abscission in lily cv. Enchantment in winter is not photosynthesis mediated, and inhibitors of ethylene production by the anthers prevent abscission even under poor light; ethylene is apparently an intermediary substance in the regulation of the abscission.

There are also several beneficial effects obtainable by treating bulbs with the ethylene generator ethephon, or the use of smoke treatments. Spraying iris plants in the field with ethephon considerably reduces the numbers of vegetative, 'three-leaf' plants. The occurrence of bud blast in the early forcing of these plants under poor light is also reduced. Disadvantages of this treatment include gum formation which makes the bulbs difficult to handle, and a shortening of the storage period because the bulbs lose weight as a result of a higher respiration rate. Alternative treatments devised in the USA involve giving freshly harvested iris bulbs a 1 h dip in a solution of $0.25–2.5 \, g \, l^{-1}$ ethephon at $20°C$ or exposing them to ethylene gas in air at $10–100$ p.p.m. for 24 h at $20°C$. After exposure the bulbs are given 3 days heat curing at $32°C$, held at $18°C$ for 2 weeks, a pre-cooling at $10°C$ for 6 weeks then grown under normal early forcing conditions. The mode of action of the ethylene is not known, although it induces cyanide resistant respiration and generally increases respiration rates in tulip, iris and freesia. High temperatures, often part of standard storage and treatment schedules of 'bulbs', raise endogenous ethylene levels, increase respiration rates and possibly affect floral initiation.

Treatment of 'bulbs' of many species, tazetta narcissus, iris and freesia with smoke or with ethephon have produced desirable results such as increased flowering percentages in borderline sized bulbs, earlier flower initiation and faster development, and replaces the need for high-temperature treatments which have long been standard procedures. Burning-over of bulb fields on the Isles of Scilly has been practised since about 1920 as a means of increasing flowering of tazetta narcissi cv. Soleil d'Or which are not lifted. A similar procedure has been in use by Japanese growers of tazetta narcissi for over 50 years. Another Japanese technique was to treat bulbs after lifting with smoke generated from smouldering wood, fresh leaves or damp straw, fed into the store. Tests have indicated that the effect of the smoke can be created with ethylene, and it is likely that the burning-over also owes its effectiveness to ethylene, although the presence of the gas in the soil has not been shown. Other 'bulb' crops responding to smoke treatment and ethylene are freesia (dormancy breaking) and iris (improved percentage flower initiation and dormancy breaking). Only small quantities of ethylene are required, a current recommendation is for 10 p.p.m. for 1–5 h daily for 4 consecutive days at a minimum temperature of $25°C$ (Imanishi, 1983).

RESPIRATION, PHOTOSYNTHESIS AND DRY-MATTER PRODUCTION

In general there is little information on these processes in 'bulbs' at the plant level, but somewhat more on crop performance and behaviour.

Respiration

Most of what is known refers to stored bulbs of tulip and iris. Experiments have shown that during long-term (355 days) storage of iris bulbs at 25°C, a loss of 2.7% of dry weight occurred. Over a 12 week period starting in early September, tulip bulbs lost 0.6% of their dry weight per week, the rates being identical at 17 and 20°C. Oxygen uptake by iris bulbs stored at 25°C was rapid immediately after lifting, but this fell by a half over 2 weeks then remained nearly constant for several months, at ca. 6 μl CO_2 output g^{-1} fresh weight h^{-1}. This rate is similar to that of stored potatoes, but lower than in most non-storage tissues. A similar drop in the respiration rate of newly lifted tulip bulbs has also been recorded to a steady state value of 10–12 μl O_2 uptake g^{-1} fresh weight h^{-1}. In iris respiratory activity increases from the steady state 'resting' condition if the temperature is lowered from 25°C. Respiration rates of excised apices of iris show a marked increase in respiration rate at the start of flower initiation, followed by a subsequent fall to the original value as differentiation proceeds, apart from a small secondary peak corresponding to the initiation of the gynoecium.

Attempts to use respiration rates or the activities of various enzymes such as catalases, peroxidases, amylases and dehydrogenases as indicators of bulb 'maturity' or developmental stages of the shoot within the bulb, especially of iris, have not been sufficiently successful to be adopted commercially.

On a field scale, respiration rates of crops have been estimated only for tulip, using shading methods to modify dry-matter production (Rees, 1967) and crop enclosures with gas analysis (Benschop, 1980). The first method gave rates in April–May of 0.04 g dry matter plant^{-1} day^{-1}, equivalent to 2.2 g m^{-2} day^{-1} on a land area basis or 20% of the gross rate of dry-matter production in full light. The gas exchange measurements showed respiration to depend on temperature, with a Q_{10} above 2 before flowering and below 2 afterwards, and rates between about 1 and 6 kg CO_2 ha^{-1}h^{-1} over the temperature range 7–30°C and three periods (2 and 1 week before flower removal and 1 week afterwards). Respiration losses of the aerial parts of the crop were ca. 16% of total gross photosynthesis in early May and 22% in June, in reasonable agreement with the shading experiment results, and with estimates for other field crops growing in spring.

Photosynthesis

Again, most of the work published on photosynthesis deals with the tulip cv. Apeldoorn, whose photosynthetic activity spans the period March–July, a total of *c*. 15 weeks. Measurements were made in The Netherlands (Benschop, 1980) using a mobile, transparent, open-system enclosure system and monitoring the CO_2 concentrations of the in-going and out-going air by infra-red gas analysis. The maximum net photosynthetic rate of 29 kg (corrected to 35 kg for complete soil cover) $CH_2O\,ha^{-1}h^{-1}$ was observed on a day in late April, when the total net photosynthesis for the day was 206 kg $CH_2O\,ha^{-1}$ and total global radiation was $159 \times 10^5\,Jm^{-2}$. Light saturation of photosynthesis was observed in June at $700–800\,Jm^{-2}s^{-1}$, but not in May with similar maximum radiation values, suggesting some loss of leaf efficiency with age despite signs of leaf senescence. Photosynthesis is independent of temperature between 10 and 20°C but decreases above this temperature, in agreement with known effects of high temperatures in shortening the growing season and reducing yield.

The photosynthetic characteristics of individual tulip leaves depends on their position. The lowest leaf (c. 50% of total plant leaf area and dry weight) has a rate of light saturation of $12.3\,kg\,CH_2O\,ha^{-1}h^{-1}$, whilst the middle leaf (c. 33% of total) has a corresponding value of 10.2. Both values are low compared with most other C3 plants. The light response curves are strongly asymptotic with a maximum rate at only a quarter to a third of full sunlight.

Removal of fully open Apeldoorn flowers just below the perianth insertion in an experiment in a day-lit controlled environment cabinet reduced the solar radiation compensation point from 110 to 60 $W\,m^2$ and increased the net rate of CO_2 uptake at 300 $W\,m^2$ by nearly 88%. As the dark respiration rates in the with and without flowers treatments were not different, it was concluded that flower removal affects net photosynthesis by increasing light interception by the leaves.

Recent work has shown the effects of defoliating Easter lily plants (Wang, 1990). The rate of photosynthesis of the fifth leaf from the apex was unaffected by removing lower leaves, and showed light saturation at a photosynthetic photon flux of 700 $\mu mol\,m^{-2}s^{-1}$. All defoliation treatments (down to 35%) markedly reduced aerial bulbil formation. Dry-matter accumulation by mother bulb scales was reduced when 50 and 75% of the lower leaves were removed, but that of the daughter bulbs was reduced only when 75% of the lower leaves were removed, indicating the respective strengths of the sinks.

Dry-matter Production

The general picture of growth as indicated by changes in dry matter is remarkably similar for tulip and narcissus, and is therefore probably representative of other spring grown 'bulb' plants. It is closely related to climate, and represents the periodicity of these plants. In general terms, as indicated by Fig. 5.8 for tulip, there is a loss of dry matter by the whole plant (initially the planted bulb) which is approximately linear until after shoot emergence in February/March, at a rate of 0.23% day^{-1}, or about 35% of the planting dry weight over the period from planting to the minimum plant weight. The old mother bulb continues to lose dry weight until it has almost completely

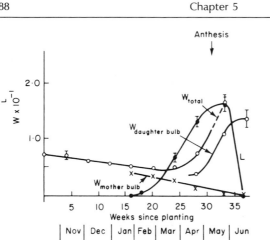

Fig. 5.8. Pattern of growth of a spring bulb as illustrated by the tulip. Total plant dry weight (W, in g) falls from planting, and the mother bulb weight continues to fall until it reaches zero. As leaf area (L, leaf area index) builds up in spring, total plant weight and that of the daughter bulbs increase, but the weight increase of the latter declines as the leaves start to senesce (Rees, 1966).

disappeared by the end of the growing season in June/July. Leaf emergence depends on situation and weather, but is initially slow until the warmer weather of April, then more rapid until maximum leaf area is attained, best described by a Gompertz curve. From this point, senescence is rapid and leaf area falls linearly to zero, giving a total growing season of about 15 weeks, depending somewhat on weather, with warm conditions allowing early emergence, and cool weather at the end of the season delaying senescence. Total plant weight increases over about 13 weeks, as there is some weight loss after first emergence of the shoot. Daughter bulb dry weight increases slowly at first, then more rapidly as leaf area reaches its maximum, then more slowly as the leaves start to senesce. The dry weight of daughter bulbs at harvest can be up to 3.4 times that of the bulb initially planted, depending on season, cultivar and the planting density adopted.

From such data the rates of net assimilation (E) and crop growth (C) can be calculated. E rises from about 2 to a peak of 9–$10 \, \mathrm{g \, m^{-2} \, day^{-1}}$ 5–7 weeks later, then falls smoothly to zero in June/July. Curves of C reach a peak of ca. $20 \, \mathrm{g \, m^{-2} \, day^{-1}}$, similar to other temperate spring growing crops, and giving values of c. 10 tonnes $\mathrm{ha^{-1} \, annum^{-1}}$. These annual values are low compared with other crops; the short growing season is responsible, and indicates the importance of delaying senescence and maintaining an active leaf canopy as late as possible.

Several authors have commented on effects on dry-matter production of differences in leaf area due to cultivar, site, season, pathogens or damage such as that caused by hail. In a comparison of four contrasting tulip cultivars grown from 9 cm bulbs, the variation in peak leaf area per plant was more than twofold, and despite emergence and final senescence dates being almost identical, the leaf area durations varied between 20.1 and 7.2 weeks. Deliberate removal of tulip leaf area to simulate hail damage has reduced subsequent yields by 10–40%. For narcissus, there are differences in emergence dates and of senescence between seasons and sites even within the UK, using identical stocks, which are reflected in yields. Rosewarne in Cornwall is an earlier site than Kirton (Lincolnshire) by 2–4 weeks, and produces a higher peak leaf area. Senescence is also earlier in Rosewarne by about 4 weeks. Similar differences can occur on a single site between seasons.

Models have been developed to simulate tulip growth and dry-matter production in the UK (Rees and Thornley, 1973) and The Netherlands (Van der Valk and Timmer, 1974). The former was an attempt to provide a simple simulation to relate temperature and light to growth and to reveal inter-relationships of growth, while the latter dealt in more detail with inter-relationships between plant density, leaf area, soil cover and rates of dry-matter production. More recently a ROCROP model of tulip growth has been developed to allow growers to estimate the effects of tactical decisions on yields, with inputs of planting bulb weight, planting density and system, cultivar specific data on leaf development and bulb grading, daily total radiation, latitude, dates of emergence and senescence, and the dates and extent of any leaf damage (Van der Valk and Van Gils, 1990).

PLANT SPACING AND COMPETITION

This is a complex subject. The optimum planting density for any given stock of bulbs depends on what end-result is required, what is being optimized, and must include a consideration of costs. In general, higher planting densities reduce the yield per plant, with the reciprocal of the lifted weight per plant increasing linearly with planting density. On a land area basis there is a curvilinear increase in total bulb weight lifted per plant, but as this is achieved as a result of increased planting weights per unit of land area, the increase in bulb weight (that is the difference between planting weight and that lifted) on a land area basis is curvilinear, usually with a well-defined optimum. However, this is satisfactory only when considering total weights. With increasing planting densities less of the total weight is represented by the larger grades of bulbs, and there are more small bulbs of less value. Plant arrangements giving best yields are those where there is an infinite array of plants, but for practical reasons there must be access, which means the adoption of beds or ridges. In the UK, narcissus are left for 2 years before lifting, so that there must be a compromise between optimum densities in the 2 years; below the optimum in the first year and above in the second. Finally, economic factors must be considered, including the relative values of the bulbs and of the land used for growing them, even if it is assumed that growing costs are independent of the planting density adopted.

Figure 5.9 shows some experimental data for tulip cv. Apeldoorn growing at five planting densities of a single grade of bulbs, in one growing season in Lincolnshire. The maximum total weight increase was at 94 bulbs m^{-2}, the weight lifted was 3.6 kg m^{-2}, and the increase in weight above that planted was 2.1 kg m^{-2}. The maximum number of forcing size bulbs (> 11 cm) was

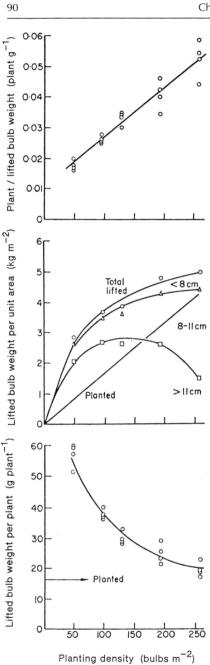

Fig. 5.9. Effect of planting density on lifted weight per plant, total yield per unit area and the yield in three component grades of bulbs of tulip cv. Apeldoorn. The yield increases per unit area (i.e. lifted weight minus that planted) is the difference between the curve marked total lifted and the straight line. (Rees and Turquand, 1969. Reproduced by courtesy of *Journal of Applied Ecology*. Blackwell Scientific Publications.)

Table 5.1. Tulip: optinum planting densities (σ, as bulbs per square metre) for different k/b ratios, where k = planting and growing costs ($£\,m^{-2}$) and b = cost of one planted bulb ($£\,bulb^{-1}$).

k/b	σ
1	15.1
5	31.3
10	41.8
20	54.6
40	69.1
60	78.2
80	84.5
100	89.3
120	92.9
140	96.1
160	98.7

Source: Rees and Briggs (1974).

$87\,m^{-2}$ at a planting density of 115 bulbs m^{-2}. Using estimates of the values of the planted and lifted bulbs by grade, the cost of the land, and the growing costs indicated that the highest profitability resulted from 65 bulbs m^{-2}. Extrapolating this information to commercial practice is beset with problems. Different grades of planting bulbs have different optimum planting densities, which cannot be resolved by using planted weight per unit area. Each cultivar has a different potential yield, the 'vigorous' ones yielding more, and capable of performing well under higher competitive stress. Differences between seasons can be greater than those between contrasting cultivars in a single season, and the values of bulbs of cultivars vary, thus upsetting the factors of the profitability estimation.

To attempt a simple synthesis for the guidance of tulip growers, a tabulated relationship was determined between optimum planting density and a ratio, k/b, where k is the planting and growing costs ($£\,m^{-2}$) and b the cost of a single planted bulb ($£\,bulb^{-1}$) (Table 5.1). It also appears that the highest profitability occurs at a peak leaf area index between 1.5 and 1.7, but this remains to be rigorously tested, although it is supported by Dutch results that dry-matter production is linearly related to the percentage soil cover.

In contrast, narcissi are planted by weight per unit land area rather than bulb number, and the analysis is simplified because weight is an appropriate criterion irrespective of the grades and types of bulbs making up the weight. For cv. Fortune in southwest England, grown for flowers and bulbs on a 2 year cycle, the highest financial return followed planting weights of between 2 and 3 $kg\,m^{-2}$. Greater precision in estimating optimum planting densities could not be achieved because of the large between-season differences observed. It is interesting that for four cultivars over 4 years, between 52 and

73% of the increase in bulb weight obtained occurred in the first year of the 2 year growing period. For Lincolnshire, where the flowers of field-grown crops are less frequently harvested, the optimum planting densities for max- imizing the weight increase per unit area of land of 2 year crops grown in ridges (ignoring any income from flowers) is lower, at $1.6–2.0\,kg\,m^{-2}$, seasonal differences being responsible for the observed range. This is in broad agreement with work in Denmark, which also stressed the cost benefit of 2 year growing of narcissus because the cost of annual lifting is greater than that of the yield increase.

Studies of the competition between weeds and crops of tulips and narcissi suggest that effects are not great on early spring growth up to flowering, but can have major effects on yield (bulb weights and grades rother than of numbers) later in the season. Weed shading of the crop effectively reduces its photosynthesis by competing for light, an effect exacerbated by the crop's earlier senescence. Competition for water can also be important in dry seasons and cause early senescence. Effects on flowering seem to be mediated through bulb size; reduced yield being associated with smaller bulbs, and hence fewer bulbs of flowering size. Perennial weeds which start into growth at the same time as the crop are most damaging, and under the worst conditions up to a third of the crop can be lost.

identical stocks, which are reflected in yields. Rosewarne in Cornwall is an earlier site than Kirton (Lincolnshire) by 2–4 weeks, and produces a higher peak leaf area. Senescence is also earlier in Rosewarne by about 4 weeks. Similar differences can occur on a single site between seasons.

Models have been developed to simulate tulip growth and dry-matter production in the UK (Rees and Thornley, 1973) and The Netherlands (Van der Valk and Timmer, 1974). The former was an attempt to provide a simple simulation to relate temperature and light to growth and to reveal inter-relationships of growth, while the latter dealt in more detail with inter-relationships between plant density, leaf area, soil cover and rates of dry-matter production. More recently a ROCROP model of tulip growth has been developed to allow growers to estimate the effects of tactical decisions on yields, with inputs of planting bulb weight, planting density and system, cultivar specific data on leaf development and bulb grading, daily total radiation, latitude, dates of emergence and senescence, and the dates and extent of any leaf damage (Van der Valk and Van Gils, 1990).

PLANT SPACING AND COMPETITION

This is a complex subject. The optimum planting density for any given stock of bulbs depends on what end-result is required, what is being optimized, and must include a consideration of costs. In general, higher planting densities reduce the yield per plant, with the reciprocal of the lifted weight per plant increasing linearly with planting density. On a land area basis there is a curvilinear increase in total bulb weight lifted per plant, but as this is achieved as a result of increased planting weights per unit of land area, the increase in bulb weight (that is the difference between planting weight and that lifted) on a land area basis is curvilinear, usually with a well-defined optimum. However, this is satisfactory only when considering total weights. With increasing planting densities less of the total weight is represented by the larger grades of bulbs, and there are more small bulbs of less value. Plant arrangements giving best yields are those where there is an infinite array of plants, but for practical reasons there must be access, which means the adoption of beds or ridges. In the UK, narcissus are left for 2 years before lifting, so that there must be a compromise between optimum densities in the 2 years; below the optimum in the first year and above in the second. Finally, economic factors must be considered, including the relative values of the bulbs and of the land used for growing them, even if it is assumed that growing costs are independent of the planting density adopted.

Figure 5.9 shows some experimental data for tulip cv. Apeldoorn growing at five planting densities of a single grade of bulbs, in one growing season in Lincolnshire. The maximum total weight increase was at 94 bulbs m^{-2}, the weight lifted was 3.6 kg m^{-2}, and the increase in weight above that planted was 2.1 kg m^{-2}. The maximum number of forcing size bulbs (> 11 cm) was

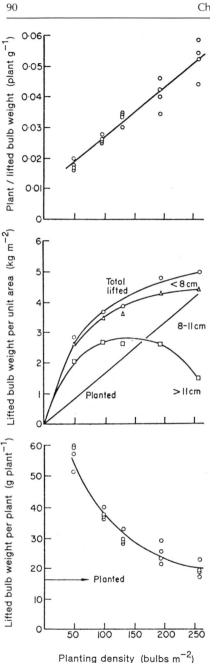

Fig. 5.9. Effect of planting density on lifted weight per plant, total yield per unit area and the yield in three component grades of bulbs of tulip cv. Apeldoorn. The yield increases per unit area (i.e. lifted weight minus that planted) is the difference between the curve marked total lifted and the straight line. (Rees and Turquand, 1969. Reproduced by courtesy of *Journal of Applied Ecology*. Blackwell Scientific Publications.)

PROPAGATION

Sympodial branching is very common in the monocotyledons, a feature that facilitates horticultural propagation by splitting off parts of the plant. Units of growth such as daughter bulbs, offsets, cormels or branched rhizomes already exist; these are capable of independent existence after the production of adventitious roots, and these units are frequently well supplied with food reserves. However, the tight programming of shoot organization in mono-cotyledonous plants (with few adventitious leaf or shoot buds) restricts the number of meristems available for propagation. In some cases this matters little – a large stock of narcissus bulbs reproduces itself rapidly enough to supply needs of bulbs for forcing, for sale and for replanting – but in other situations, such as when there is a need to increase rapidly the numbers of an attractive mutation, a new cultivar, or of small numbers of virus-free 'bulbs', it is clearly advantageous to have a more rapid, artificial means of multiplication.

Natural rates of multiplication depend on genera and species. In narcissus it is slow ($\times 1.6$ annum^{-1}) but in other species with bulbs, more daughter bulbs are formed, often at least one in the axil of each scale as in tulip and iris, giving a rate of about $\times 5$ annum^{-1}. Many lily species produce aerial bulbils in the leaf axils; these can number several dozen from a large plant (Fig. 6.1). Large numbers of cormels (strictly tubers) are also produced by gladioli, forming clusters on short outgrowths or stolons between the mother and daughter corms. A hundred or more cormels can be produced by a single large gladiolus plant. The need for artificial means of propagation is therefore largely dictated by the natural rate.

Because of the range of plants within the scope of this book, it is not surprising that there are many methods by which they can be propagated artificially. Natural multiplication is often slow, a reflection of the morphology of the plants, and their tendency to produce propagules of some size, in line with the general survival strategy of the group towards investment in small numbers of offspring, well provided with food reserves. When large numbers of propagules occur, they are small, and it can take several years of growth in a juvenile, non-flowering form before they reach the mother 'bulb' size.

In the wild, vegetative multiplication alone is not satisfactory; often producing clumps of plants derived from a single individual, which leads to severe competition for light, nutrients and water, poor growth and the loss of flowering capacity in the central plants (as also observed in the garden situation). They are also more susceptible to pest and disease attack because the plants are a clone, genetically identical and therefore of equal susceptibility to a pathogen. In some cases these shortcomings are partly overcome by the dispersal mechanisms evolved for ensuring that the propagules are established some distance from the parent plant. These include the droppers of tulip, and several forms of contractile tissue which can draw growing points away from the parent 'bulb', often in ingenious ways, along specialized tubular structures (see Fig. 4.10). Most frequently dispersal is a property of seeds.

Whilst vegetative propagation is almost universal in the horticultural propagation of 'bulb' plants, in the wild there is much greater dependence on seed propagation. There are several reasons for this, probably the most important being the horticultural need to preserve the clonal nature and uniform characteristics of each cultivar; use of seeds would introduce unwanted variability. Further, as seen in Chapter 3, many cultivars are sterile and do not set seed. Once a grower has built up a reasonably large population of a cultivar, natural rates of multiplication are often adequate, with little or no extra input required for artificially increasing numbers. In situations where it is desirable or advantageous to propagate rapidly, vegetative methods are often available, which preserve the clonal nature of the offspring, usually involving dissection of some parts of the plant to give large numbers of units, which are then grown on to produce new plants, either by fairly crude methods such as growing individual lily scales, or by the sophistication of micropropagation *in vitro* under sterile conditions.

Little is known about the relative effectiveness of vegetative and seed propagation in maintaining wild populations of 'bulb' plants. For our native bluebell, *Scilla non-scripta*, it is believed that populations are maintained almost entirely by seed, the contribution of vegetative multiplication being estimated as only 4%. In contrast, in wild *Narcissus pseudonarcissus*, clonal growth has been shown to be much greater than the contribution of seed, and individual plants and their vegetative offspring have a half life of 12–18 years.

SEED PROPAGATION

Whilst propagation from seed is not widely used commercially, except for some individual crops, such as anemone, it is vital for the breeder concerned to produce new cultivars of species normally propagated vegetatively. Whilst

10

Fig. 6.1. Stem bulbils of lily 'Enchantment' which are produced in the axils of leaves on the flowering stem. They have only a few scales, and often show roots by the time they become detached from the parent plant. Scale in mm.

some genera and species readily produce seed in the field under normal northern European conditions, others respond to protection from rain when the seed is developing and maturing, and much commercial breeding is in unheated glasshouses. It might also be necessary to pollinate the flowers artificially if there is a shortage of pollinating insects at flowering time because of bad weather, especially cold. For narcissus, there seems to be little evidence of pollination in the absence of insect pollinators, which, in the UK, appear to be a number of species of bumble bee of the genus *Bombus*. Freesias require warm conditions and pollinators; it is usual to produce F_2 seeds in heated glasshouses equipped with beehives. For F_1 progenies, better pollination results from hand crosses because each plant is genetically different. A curious case is that of *Fritillaria imperialis* whose large flowers are pollinated by blue tits, the only known case of bird pollination in the UK. Field studies in Israel have shown that pollination by beetles is common in anemone, ranunculus and *Tulipa agenensis*. Some bulbous plants spread rapidly in the garden situation by seeding. Bluebells are notorious for this; other genera which spread by seed are snowdrop, muscari and crocus.

The seed of some 'bulb' plants has limited viability, so it is advisable to plant it when fresh. Further, many spring flowering species also have a cold requirement before germination, which is satisfied by overwintering the planted seed in cold frames, or similar, to produce seedlings in spring. There is little critical information on optimum temperatures and durations to ensure rapid germination, on narcissus, for instance, because it is easy to saturate the requirement by using natural winter cold. A duration of 12 weeks at 12°C has

proved effective for a range of commercial narcissus species. Early germination by adopting the optimum temperatures can be counter-productive
because the resulting seedlings emerge out of season and grow slowly. For
tulip, optimum temperatures are in the region of 4–10°C, for 6–8 weeks, with
different optima with different species, and even with seed batches of the same
species. Three species of *Chionodoxa* and *Muscari armeniacum* germinate
within the range 1–9°C, with an optimum at 5–7°C, the mean time at the
optimum being 92–117 days. Other species have different requirements;
Narcissus bulbocodium, which comes from hot, dry parts of the Iberian penin-
sula and northwest Africa, can be induced to germinate rapidly and synchro-
nously at 5–16°C by prior treatment at 26°C. This is interpreted as a mechanism preventing germination in summer. Hyacinth seed germinates best at
17–20°C. Many large-seeded members of the Amaryllidaceae, such as *Amaryllis, Crinum, Haemanthus, Clivia* and *Nerine* have an outer seed covering
which restricts water loss, so that, with no specific temperature requirement,
these seeds germinate 'dry' in the seed packet, and snowdrop seed germinates
viviparously.

Germination is of two kinds, hypogeal and epigeal (Fig. 6.2). In the
former, the cotyledon remains underground, and the first true leaf, which
may be tubular or spatulate, emerges from the side of the cotyledon and
becomes aerial. Emergence of this leaf may be delayed for several months,
from a small underground bulb and its attached root. In the latter, the
cotyledon is green and leaf-like, often hooked-over, with the seed at its tip, and
not usually accompanied by a leaf in the first year. Narcissus seeds are
hypogeal, whilst those of tulips are epigeal. Lily seeds can be either, depending
on species. A long seed dormancy allied to hypogeal germination with late leaf
emergence can delay very considerably the above-ground appearance of
seedlings.

Some plants have an extended period of juvenility following germination,
others can flower in the same year as the seed is planted. Those flowering
rapidly include *Tigridia pavonia* and *Lilium formosanum*, but some tulip species
can, exceptionally, take up to 7 years. A delay of 3–5 years is more usual.
During this period, the initial daughter bulb either increases in size if it is of
a kind that persists for more than one season, like narcissus, or produces
increasing numbers of small bulbs annually if it has a non-persistent bulb like
tulip. Leaf numbers are low in plants with a long juvenile period; until
flowering a tulip always has only one leaf (Fig. 6.3), narcissi have one or
sometimes two in the first 2 years, increasing to three or more only in the
fifth year.

Apart from the requirements for breeding, seed multiplication is therefore
used only where there are advantages of large and reliable seed set, associated
with short juvenile periods, and no requirement to avoid the greater variability introduced into the stock compared with vegetative reproduction. The
only other advantage is that in almost all cases, plants produced from seed are

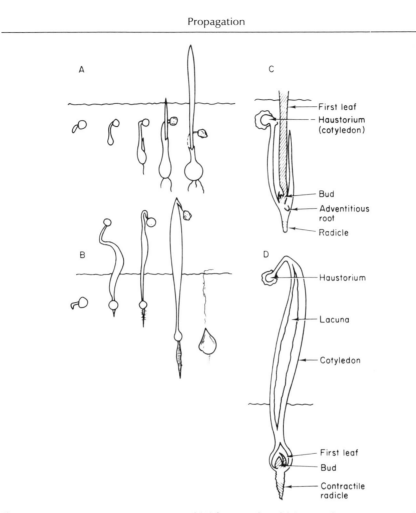

A

C

First leaf

Haustorium
(cotyledon)

Bud

Adventitious
root

Radicle

B

D

Haustorium

Lacuna

Cotyledon

First leaf

Bud

Contractile
radicle

Fig. 6.2. Diagrammatic representation of (A) hypogeal and (B) epigeal germination with sections of (C) hypogeal and (D) epigeal seedlings (Rees, 1972).

free of virus infection – unlike vegetatively propagated material. The following crops are commonly propagated by seed: *Anemone* (see Chapter 7), *Allium*, *Begonia*, *Chionodoxa*, *Cyclamen*, *Eranthis*, *Freesia*, *Fritillaria*, *Liatris*, *Puschkinia*, *Sparaxis*, *Tigridia* and *Ranunculus*.

Cyclamen persicum, the commonly grown pot plant, can be used as an example of a seed-raised crop. Pollination is done by hand, transfer being facilitated using a small brush or pipe cleaner. After fertilization, the peduncle elongates, lowering the pod close to the soil. Seed takes 2–3 months to ripen, and there are c. $100 g^{-1}$. It is collected and stored at 15–20°C in dry conditions, but if necessary can be kept for several years at 2–10°C without serious loss of viability. Germination is straightforward, with no requirement

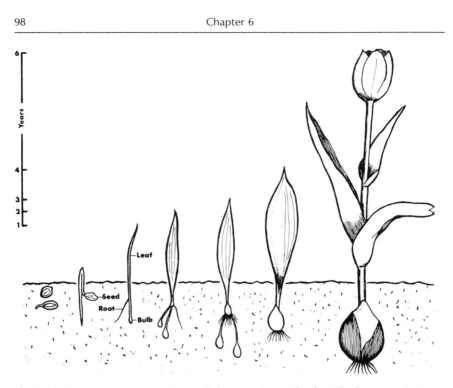

Fig. 6.3. Diagrammatic representation of the annual growth of a tulip from germination to flowering. Note the single leaf until flowering in the fifth year.

for after-ripening or for special environmental conditions. Germination is most rapid at 15 or 20°C in the dark, and this is usually done in trays with seed spaced 7 × 7 cm apart and 6 mm deep in coarse peat limed to a pH of 6.0. A temperature controlled room is used rather than a greenhouse because higher temperatures (above 22°C) can be inhibitory. Germination is rapid, roots emerge in 5 days, with hypocotyls visible in *c.* 28 days. When the cotyledon petioles start to extend, the seedlings should be transferred to a greenhouse to grow six or seven unfolded leaves, when they are potted. A feature of germination in this species is that a large proportion of the seed reserves are used in the formation of the 'corm'; the hypocotyl begins to swell immediately after the primary root has emerged. Expansion of the photosynthetic area is therefore slow, and this contributes to the long vegetative period. For more details see Widmer (1980).

PROPAGATION BY CUTTINGS

'Cuttings' is used here in a broad sense to apply to any artificial division of a plant into parts which can then regenerate a new plant. Many plants with a large, amorphous rootstock, often rather undefined morphologically, can be roughly divided using a spade or similar implement, into several smaller units for successful replanting. Similar but less crude is the normal method for propagating alstroemeria by lifting the plant clumps in summer and separating the tangle of rhizomes. From these the apex and up to 10 cm of healthy rhizome is removed with roots and tubers attached, and the aerial shoots trimmed back. The resulting 'split' is grown in a small pot to become established before transplanting. Similar procedures are used to prepare 'crowns' of *Convallaria*.

Some 'bulb' plants have the capacity to regenerate plantlets from leaf cuttings. Mature leaves removed from the growing plant are used entire or cut into several pieces, and placed with the cut ends (after dipping in hormone rooting powder) in sterile rooting medium in a seed tray. Alternatively, leaves can be laid on the surface of the medium and pinned down with wire loops to maintain good contact with the medium. Cutting across prominent veins can help induce plantlet formation on the wounds. It is essential to prevent the leaves from drying out, by enclosing the tray in a plastic bag, and it is usual practice to provide warm conditions and shade. Small plantlets form within a few weeks. After roots are formed, they can be transplanted. The method has been used successfully for *Haemanthus, Hyacinthus, Lachenalia, Lilium, Muscari* and many aroids. It could probably be applied to many others.

Many 'bulb' plants have no elongated stem which can be cut into pieces to root, although this procedure can be used for dahlia, canna, some begonias and lilies, stem cuttings of which, taken after flowering, will produce bulbils in the leaf axils. Although not used much as a commercial technique, it is probably true that, given care to avoid infection, the corm of any plant can be cut vertically into several pieces, each of which will grow if it has a pre-existing growing point, or is capable of generating an adventitious one. Similarly, stem tubers can be cut, as can root tubers provided that each tuber has a portion of attached stem with a bud. The basic requirement is for a bud, food reserves, the capacity for rooting and the prevention of fungal or bacterial attack of the cut surfaces. Many mature bulbs can be cut vertically into sectors with the scales held together by a part of the base plate to produce propagules which can then be incubated in a rooting medium. Such basic techniques have been refined for narcissus and other amaryllids as 'chipping' and 'twin-scaling'. It is most convenient to cut 'bulbs' when they are 'dormant', and where experimental evidence exists, productivity is highest then; for example, narcissus twin-scaling or chipping is best done in June–September, and hippeastrum in November–January, although some of the easier subjects, like *Nerine* and *Hymenocallis*, can be treated at any time.

Scaling, Scooping, etc.

A range of treatments can be used on responsive genera to produce

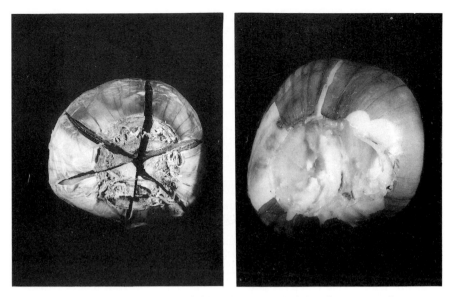

Fig. 6.4. Bulbs showing cross-cutting (left) and scooping (right) techniques used to induce bulbil formation on the cut surface of the base plate after destroying apical dominance. Both bulbs viewed from below. This technique is used commercially on hyacinth bulbs.

adventitious buds, which can then be grown on to adult plants. All involve cutting out the main growing point of the bud, thereby removing the effect of apical dominance which inhibits the initiation and development of adventitious buds. In most cases the increase in numbers which can be achieved is high, in comparison with either natural increases or those considered so far. Some are long established techniques, like the cutting and scooping of hyacinths and the scaling of lilies; more recently developed for commercial use are the twin-scaling and chipping of narcissus and other species. The fact that a technique will work does not necessarily mean that it has commercial potential. Another method might be superior in terms of rates of multiplication, or be cheaper relative to numbers produced. Thus, leaf cuttings of muscari and hyacinths are possible means of propagation which are never used commercially.

Cross-cutting and scooping

Scooping was developed in the 19th century from the old system of growing selected hyacinth bulbs to a large size when there was some loss of apical dominance and a tendency to form clusters of daughter bulbs (Fig. 6.4). Wounding the base of the bulb increases the bulbil numbers, so this is deliberately done using a curved or spoon shaped knife to remove the base plate leaving a shallow depression in the middle of the bulb. The depth of the

Fig. 6.5. Bulbil formation on detached bulb scales of lily incubated in vermiculite for 4 weeks at 23°C. Note the long roots.

cut is important; the base plate must be removed without damage to the cells at the base of the scales which retain the capacity for regenerating new meristems. Cross-cutting, first used in about 1935, is a less skilful operation which involves cutting into the base plate from below with three or four incisions across the diameter of the bulb sufficiently deep to damage the shoot fatally (Fig. 6.4). In both methods, after treatment with fungicide, the bulbs are kept for a week at 25°C in a dry atmosphere to encourage callusing, then at *c.* 90% relative humidity (r.h.) as the bases of the scales swell. After 3 months the bulbs are hardened off and planted in early December. About 25 bulbils are produced from each mother bulb, depending on cultivar and bulb size. After growing on for 3 years, about 70% achieve 12 cm grade or larger. Compared with scooping, fewer bulbils result from cross-cutting, but they are larger and are consequently saleable a year sooner. The technique is also used for *Scilla sibirica* and *Fritillaria imperialis*.

Scaling

This is a long established technique used for lilies, based on the knowledge, dating from at least the 14th century, that each scale detached from a bulb and planted, will produce a new plant (Fig. 6.5). Scales are broken away from the bulbs and, following a disinfectant dip, are planted in sterilized rooting medium or in a thin walled polythene bag containing moist vermiculite (five parts dry vermiculite to one of water). Incubation temperature and

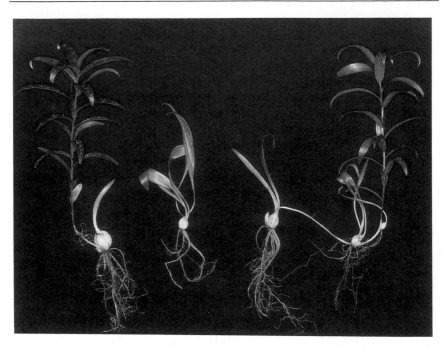

Fig. 6.6. Plantlets of lily 'Enchantment' produced from scales. Note that the plant on the left has an emergent leafy stem, whilst its neighbour has only radical leaves. The third plant has horizontally growing stems with additional bulblets.

duration depend on species or cultivar, but are typically 24°C for 8 weeks followed by 17°C for 3 weeks. Before planting in sterilized compost or soil, the scales are separated from the vermiculite and dusted with a systemic fungicide. Recent investigations on Easter lily have shown that hydrolysis and mobilization of reserves first occurs at the tip of the detached scale, to support growth of the bulbils at the base of the scale. Scaling can also be used for hyacinths, some fritillaries, muscari and scillas. It is a well established large-scale commercial technique in the USA to break scales away from selected Easter lily plants in the field shortly after flowering or at bulb lifting time in the autumn, and scatter them in rows or furrows in a warm, well drained site (Fig. 6.6). The bulbils, which are initiated on the scales, sprout the following spring.

Twin-scaling

This is an old technique, known since 1935 and described in *The American Gardener's Book of Bulbs* (Everett, 1954) as 'stem cuttage' of hippeastrum. Cutting the 'dormant' parent bulb vertically into several wedge shaped segments, and further cutting the wedges between the scales produces a

Fig. 6.7. Narcissus twin-scales after incubation, with the bulbils which form in the axil between the scales.

number of pieces, each comprising a few scales and a piece of the base plate. These develop bulbils when planted in moist compost (Fig. 6.7). More recently this technique has been revived and refined for use with narcissus and some other bulbs. The technique was used at the Glasshouse Crops Research Institute in the early 1960s for propagating virus-free narcissi, and was improved by Alkema at Lisse, The Netherlands, at about the same time. He refined the method of cutting the bulb into two adjacent scales linked by a piece of base plate, each portion being called a twin-scale, and the process twin-scaling. A further improvement was to incubate the twin-scales in moist vermiculite in polythene bags at controlled temperature for the bulbils to develop before planting.

The currently used method is as follows. A 'dormant', flowering size, parent bulb is first checked for trueness-to-type and health. It is cleaned of old roots, the outer parts of the base plate and old brown scales, the top half of the bulb is cut away and discarded, and the bulb dipped for a couple of minutes in a sterilizing solution. The bulb is then inverted and cut vertically with a large sterilized knife into eight or 16 wedge shaped segments, each with a piece of the basal plate. Further cutting with a sterilized scalpel separates the scales in pairs, with the two scale portions of each pair linked by a piece of base plate (the whole forming the twin-scale). A large bulb will provide 60–100 such twin-scales, from which 48–90 bulbils will result after appropriate incubation.

For reasons not fully understood, bulbils form in the majority (over 80%) of narcissus twin-scales, provided they are not too small, whereas in single scales with a portion of base plate they rarely (2–7%) do so. It is hypothesized that with twin-scales, as opposed to single scales, there is a band of un-damaged totipotent cells at the base of the scales which can differentiate to produce the adventitious buds. In hippeastrum, single scales first produce protocorm-like bodies which then develop bulbils, whereas twin-scales produce bulbils directly. The new bulbil is initiated on the abaxial surface of the inner of the two scales, but develops vascular connections with the outer scale. The rate of bulbil development is controlled by the length and thickness of the outer scale.

Much experimental work has gone into further refinements of the method for narcissus (for a full account see Hanks and Jones, 1986). It is essential that the whole operation is kept as sterile as possible by washing down working surfaces and keeping instruments sterilized periodically whilst working. It is also necessary to protect the newly cut twin-scales from drying out and fungal attack before they are bagged. To this end they are treated with a soak in a high concentration of fungicidal solution or dusted with fungicidal powder before being incubated in polythene bags of moist vermiculite. Attention to detail is necessary; the thickness of the polythene bags, the wetness of the vermiculite, the proportion of twin-scales to vermiculite, the air space above the scales and the composition of the incubation medium all affect the final success rate. The duration of incubation and the temperature are also import-ant; the optimum depends on cultivar, but an indication of the successful range is 15–22°C for 12–16 weeks, in the dark. If prolonged, the bulbils begin to shrivel. Ideally, at the end of the incubation period, at planting, the scales should be depleted of reserves but there should not be excessive shoot and root growth. Other incubating media have been tried, but none is more successful than vermiculite.

There are interesting differences between the various kinds and ages of twin-scales. Outer twin-scales are thinner, contain less food reserves, but produce bulbils faster than the inner, fatter twin-scales, whilst those with a scape base as half the twin-scale perform badly. The size of twin-scales cut is also important, as the success rate decreases as their size is reduced. The best time to twin-scale is July/August. As a working guide, a 12–14 cm bulb should be cut into eight segments, giving 32–40 twin-scales. It is wasteful to cut the twin-scales too small, because of the loss to the overall efficiency of the system of the food reserves in those twin-scales that produce no bulbils. Larger twin-scales result in fewer, but larger bulbils, so the choice of number (size) of twin-scales cut from the mother bulb depends on the competing demands of large numbers of plants later, or a smaller number earlier (Table 6.1). The ideal is the minimum twin-scale size (weight) that will initiate and support the growth of a single bulbil, and is 0.5–2.0 g, depending on cultivar. Care is required in subsequently growing the bulbils after their removal from the

Table 6.1. Effect of size of twin-scales on bulbil formation. *Narcissus* cv. Fortune bulbs of 12–14 cm grade were cut into 8–32 segments, and each segment cut into four twin-scales. The number and weights of bulbils were determined after a year's growth. Note that with increased numbers of twin-scales cut, bulbil number increases, steadies, then falls, whilst bulbil weight per initial mother bulb falls regularly. The choice lies between the 8-cut and the 16-cut, depending on the relative value placed on bulbil number and weight.

	Yield per bulb	
Cutting method	Bulbil no.	Bulbil wt (g)
8 segments, 32 twin-scales	31	44
16 segments, 64 twin-scales	47	30
24 segments, 96 twin-scales	45	19
32 segments, 128 twin-scales	18	10

Source: Hanks and Jones (1986).

incubation medium, and this is usually done in frost-free polythene houses or glasshouses in sterilized, prepared beds for 2 years before the bulbs are large enough to be treated like a normal stock. From cutting, the bulbs should reach flowering size in 3–4 years.

Twin-scaling has been used successfully for many other genera, including *Allium, Chionodoxa, Fritillaria, Galanthus, Haemanthus, Hippeastrum, Hyacinthus, Hymenocallis, Iris, Lachenalia, Leucojum, Muscari, Nerine, Ornithogalum, Pancratium, Scilla, Sternbergia* and *Veltheimia*. Some are easier than narcissus (*Hippeastrum*), others more difficult or more slow (*Nerine*), or require special treatment (*Iris* bulbs must be cut first in the plane through existing daughter bulbs). There is no gain over natural multiplication rates of tulip, whilst hyacinth performs better when cut into single scales rather than twin-scales.

Chipping

This is a method derived from twin-scaling which is less critical, and cruder, but can be mechanized and used on large batches of bulbs. It has been described as twin-scaling simplified, although it is closer to the original stem cuttage. Basically it is the first part of the procedure described above for twin-scaling. The bulbs, preferably round, 10–12 cm grade, each weighing ca. 35 g, are cleaned of dry outer scales and surface sterilized before cutting into 16 radial segments (chips) with a sharp, sterilized knife (Fig. 6.8). The ideal weight for a chip is about 2 g. After a fungicidal dip, the chips are 'planted' in alternate layers with damp vermiculite in shallow trays enclosed in thin gauge plastic bags and kept at 20°C for 12–16 weeks. The chips, with

5 cm

Fig. 6.8. Narcissus chips at the end of incubation showing bulbils. (Photograph courtesy *Horticulture Research International*, the copyright holder.)

the attached bulbils, are then removed from the trays and grown on in a frost-free polythene tunnel for 2 years. Some will flower in 2 years from chipping, and most by the third year. The bulbs can then be lifted and treated like a commercial field-grown crop, or can be chipped again.

Machines developed for chipping have either static or spinning blades; they can cut a trimmed bulb into its 16 chips with one movement, so that a throughput of several hundred bulbs an hour or 0.5 tonne day^{-1} is easily achieved. Because chipping results in large numbers of propagules, simpler handling methods have been investigated, including direct planting and planting in nets to aid recovery. For incubated chips, chipping in June–October, followed by planting in October–December, gives good results, but for those planted directly into soil, chipping and planting in June and July are best. Later planting gives poor yields because the small plants are exposed to low temperatures, below the optimum for bulbil production. Chips planted directly into the soil are positioned 7.5–10 cm deep; if shallower, the resulting bulbs are long and thin. Although planting can be done by machine, the angular shape of the chips impedes their flow from a hopper, so that hand planting is usually better. A planting density of 100 cm^2 per chip gives maximum yield, but the spacings adopted are frequently less generous. This is particularly so when expensive growing facilities are being used – such as a polythene tunnel (which can increase yield by 20% above that in the open), or an insect-free house and sterile compost to safeguard the bulbs' virus-free status. Bulbs are normally left for 2 or 3 years before lifting.

Current general recommendations for chipping are to cut bulbs into eight segments which are then incubated or planted directly, preferably in July at a rate of 2–5 tonnes of chips ha^{-1} in ridges in the field. Chipping is important for cultivars like Tête-a-Tête, which are increasing in popularity and therefore in short supply; most bulbs of this cultivar are produced by chipping.

MICROPROPAGATION

This is a general term for the vegetative multiplication of plants by tissue culture, *in vitro*. It is considerably more sophisticated than any of the methods described above, and requires capital investment in equipment, high running costs and expertise. However, these disadvantages are balanced by high rates of multiplication, making it worthwhile for certain crops or for some situations, where very large numbers are required quickly, and economies of scale reduce the unit costs to acceptably low values. As a general propagation method (in contrast to specialized, small-scale activities for special purposes) micropropagation is not widely used for 'bulb' plants. Recent figures show that of the 60 million plants produced in The Netherlands annually by micropropagation, 13 million were 'bulbs', of which 12.6 million were lilies.

Basically the technique consists of growing plant cells or parts, called explants, on a culture medium in culture vessels (usually small plastic or glass screw-top jars), under controlled temperature and light, and manipulating the growth of tissue and the initiation and development of organs by adding various hormones and growth factors (Fig. 6.9). To avoid contamination by bacteria and fungi, the flasks, culture media and explants are sterilized, and all handling of cultures is done under strictly aseptic conditions in laminar airflow cabinets. Media consist of salts providing essential elements, a sugar source of carbon and the vitamin thiamin. The most commonly used medium is that of Murashige and Skoog, often modified slightly to suit the special requirements of the species being grown, and solidified with agar to support the plant material. Initiation of shoots and roots in culture depends on the balance of cytokinins and auxins in the medium; a high cytokinin:auxin ratio favours shoot formation, and a low one favours roots.

For setting up a production system, the appropriate explants must be identified. Although all plants have a potential for regeneration from some part, it is necessary to select tissues or organs which respond well; these are usually meristematic regions with actively dividing cells, such as shoot apices, axillary meristems in leaf axils, and the meristems at the bases of linear monocotyledonous leaves. For bulb plants, the most successful explants have been scales, generally from the basal part near the junction with the base plate, stems, especially thin slices from young, unelongated stems, and

Fig. 6.9. Culture jar of micropropagated narcissus plantlets induced to form bulbs prior to transfer to growing *in vivo*. (Photograph courtesy *Horticulture Research International*, the copyright holder.)

already differentiated buds. There are three main methods of tissue culture propagation: the culture of shoot tips and proliferation by axillary shoots, the induction of adventitious shoots on organ explants, and the production of adventitious shoots from callus (Hussey, 1978).

Gladiolus can be rapidly propagated by axillary shoot proliferation, starting from explants of axillary buds, with a rate of multiplication of × 3–5 every 6–8 weeks by serial subculture. Iris daughter bulbs can be excised and cultured, or explants of 1 mm thick stem sections can be used to produce numerous adventitious shoots; from a single bulb, 100 adventitious shoots can be produced in 12 weeks. Freesias are micropropagated by axillary shoots, produced from explants of corm tissue or stem sections just below unopened flower buds. Lilies are readily micropropagated from stem sections or bulb tissue, with rates of × 50–100 within 6–8 months. With carefully designed mass production systems, considerably higher rates are claimed to be possible, up to 1.2×10^{10} for *Lilium speciosum* and 3.2×10^{12} for *L. auratum* in 1 year.

There are several possible starting points for narcissus micropropagation: explants of leaf bases, the lower 2–3 mm of scape with some base plate tissue, or twin-scales have all been used successfully. High levels of auxin result in a callus-like growth with a capacity for continuous shoot differentiation on transfer to a medium high in 6-benzylaminopurine. Strong apical dominance prevents all but the most vigorous shoots from producing axillary shoots, but splitting these shoots vertically regenerates secondary adventitious shoots, so that from a single mother bulb, 1000 bulbils can be produced in 12–15 months. Many other genera have been micropropagated successfully, and easily, including *Nerine, Hippeastrum, Allium, Ranunculus, Shizostylis, Sparaxis* and *Agapanthus*. Others, less amenable to rapid propagation (although this is changing as more appropriate selection of explants and better culture methods are being developed), include *Fritillaria, Galanthus, Ipheion, Lachenalia* and *Muscari*. Despite considerable effort, only limited success has been achieved in the micropropagation of tulip.

A critical phase in micropropagation is the transfer of plantlets from the culture vessel to soil or other growing medium. If this involves an active green plant, root production must first be encouraged by transferring the growing culture to a medium with low or no cytokinin, with or without a high concentration of indole-3-butyric acid. After removal from the culture vessel to a sterile growing medium, high humidity using mist or plastic cover around the plants will help them become established. An attractive alternative is to allow or encourage the plantlet to become 'dormant', as a small 'bulb', by a high sucrose concentration in the medium. Before planting such 'bulbs' it might be necessary to break their 'dormancy' with low temperature (0–9°C for 2–10 weeks, effective with gladioli, lilies and narcissi), or high temperature (45°C for 1 h with *Lilium longiflorum*).

A concern with micropropagation is the genetic stability of the cultured material. All vegetatively propagated plants are subject to mutation, usually at a low rate depending upon the methods used. Adventitious shoot meristems are more prone to mutation than axillary shoot meristems because they are more frequently derived from single cells. Callus shows a strong tendency to polyploidy or mutation, and any system involving callus must be rigorously tested for genetic stability; whilst some callus cultures have proved to be remarkably stable genetically, others have proved to be less so. With such rapid rates of propagation, it is easy to produce large numbers of plants which are genetically useless. Frequent checking is necessary to avoid having to discard large numbers at a late stage. Unfortunately, a complete check has to await flowering, which occurs near the end of the process.

MULTIPLICATION RATES

The natural rate of increase in narcissus numbers is low: from one bulb to increase to 1000, assuming no disease or pest problems and good husbandry, takes about 16 years, an annual multiplication rate of × 1.6. For twin-scaling, if we make the reasonable assumptions of 36 bulbils per bulb, allowing 4 years for a bulbil to grow to the size of the mother bulb to repeat the procedure but with no losses along the way, this method can result after four cycles of cutting and growing, in 1.7 million bulbs the same size as the mother bulb. A rate of × 36 every 4 years is equivalent to an average increase of × 2.45. This is comparable with chipping, where a lower rate of bulbil production (16) from each cutting operation (one bulbil from each of 16 chips per mother bulb) is compensated by a shorter time for these bulbils to grow to the size of the mother bulb (2 years instead of 4). The advantages of chipping are its greater simplicity and possibilities for mechanization.

Micropropagation is still more rapid, with 500–2000 units available from an initial narcissus bulb in 18 months. Put another way, rates of × 27 annum^{-1} are possible, which is equivalent to 1.6 years from 1 to 1000. To this must of course be added a period to allow the propagules to grow to the size of the original mother bulb, perhaps a total of 5 years to compare with the other estimates above, and an overall multiplication rate of between × 3.5 and × 4.5. Care must be exercised not to calculate over-enthusiastic rates of multiplication based on the best results obtained with the most responsive cultivars in experimental work, where conditions are near ideal, and expect these to be matched regularly in production line practice, where substantial losses occur for many reasons.

In an assessment of narcissus micropropagation using semi-commercial methods, Squires and Langton (1990) concluded that an average narcissus cultivar might reasonably be expected to yield c. 1200 flowering sized bulbs from one initial bulb in 4–5 years. The time taken to produce each bulbil, whether viable or not, was 2.6 operator minutes, an indication of cost, although a fully commercial production line could probably reduce this considerably. The authors concluded that, for narcissus, micropropagation is not an attractive alternative to chipping, and is best regarded as a complementary process for use with small stocks of elite material.

In a programme of multiplication involving large numbers of 'bulbs', a combination of several methods would be advisable. Micropropagation is ideal to achieve an initial high number, but it is too expensive to continue into large numbers. Depending on facilities available, a second stage could involve either chipping, or twin-scaling followed by chipping, until the numbers became too large, when recourse to natural rates of multiplication with optimization of all growing factors to maximize yield per plant would be the only feasible treatment for the main bulb of the stock. The exact population numbers at which a switch to slower methods should be made cannot be generally defined, as they depend on the individual circumstances, the target population required and its timing.

7

BULB, CORM AND TUBER PRODUCTION

A major part of the industry is concerned with the production of bulbs, corms and tubers which are then used for several purposes. They may be sold to gardeners, local authorities, parks departments and the like for outdoor planting, they may be grown for cut-flower production outdoors or under glass or other protection, or be used for pot-plant production. Sometimes the grower grows the bulbs, then forces them and sells the resulting flowers, thereby maximizing financial returns.

The production of 'bulbs' usually involves natural multiplication of plant numbers, so that different grades are produced with different potentials, allowing some 'bulbs', usually the larger ones, to be sold or forced whilst still retaining sufficient of the smaller grades as planting stock for continuing the process. In some cases, use is made of artificial methods of propagation which results in high rates of multiplication, but which usually require the propagules to be grown for several seasons to achieve commercially valuable, flowering sized material, as seen in the previous chapter. Grading of 'bulbs' is an essential part of growing to separate the lifted crop into sizes suitable for sale for several purposes, for growing on in the following season, as well as to ensure continuity of supply of replanting material. There is the possibility of recovering some forced 'bulbs' after they have flowered; it is common practice to replant narcissus bulbs in the field, where they remain for 2 years, but not tulips, whose forced bulbs are discarded. The decision is an economic one, as tulips take too long to reach saleable size to justify the costs of two seasons of planting, husbandry, lifting, handling, etc. When bulb prices are low, some growers consider it uneconomic to recover narcissus bulbs.

The range of crops precludes giving detailed information on each one, especially as there are several aspects of culture which are common to many of the major crops. More details are given for narcissus, with indications of how other cultivations differ. Further information is available in the Ministry of Agriculture, Fisheries and Food (MAFF) publications on narcissus and tulip

Fig. 7.1. Early narcissus growing in Cornwall. Note shelter-belt hedge and trees, the stony soil and the sloping ground.

bulb production (MAFF, 1982a,b, 1985), and on bulb and corm production (MAFF, 1984b).

NARCISSUS BULB PRODUCTION

The choice of site for commercial narcissus growing depends on local conditions of climate and soil. Within the UK narcissi can be grown successfully almost anywhere where other crops are grown. Of the total UK area of *c.* 4500 ha most (*c.* 70%) in the eastern counties from the Humber down to Norfolk, with the greatest concentration in south Lincolnshire. The second important area (*c.* 18%) is in the Southwest, mostly in the extreme west of Cornwall, on the Isles of Scilly and in the Tamar Valley of Devon and Cornwall; this area is expanding annually by about 1% of the national total at the expense of the eastern counties. The remainder are in Scotland (*c.* 6%), Northern Ireland, Wiltshire and Pembrokeshire, although the majority of counties in England and Wales have some commercial crops, often of very small area. Current agricultural diversification is tending to increase the amount grown in the non-traditional areas.

Even within the UK, climatic effects are important; the mild climate of the Southwest allows early, good quality outdoor flower production (Fig. 7.1), with bulb production as a less important aspect of cultivation, in contrast to the eastern counties where flowers are picked only if the market is favourable,

Fig. 7.2. A large field of narcissus in Lincolnshire. The silty soil is stone-free and flat, and the fields are large; ideal for large-scale mechanized growing.

and the chief aim is bulb production, allied to subsequent flower forcing (Fig. 7.2). It is also essential that there is sufficient moisture from rainfall and a high soil water holding capacity (a minimum of 40 mm in the top 300 mm) during the growing season, especially between flowering and senescence in about June.

It is important to consider plant health factors and rotations to avoid disease carry-over. For narcissus, freedom of both planting stock and soil from narcissus stem nematode is essential, and host plants of this pest (like onion) should not be included in the rotation, which should ideally include narcissus not more frequently than once every 7 years, with a minimum of 4 years between bulb crops. Land availability to allow such rotations can often limit narcissus growing. Whilst narcissi are fairly tolerant of soil conditions, they do best at a pH of 6.5 or above, with a minimum organic matter content of 3%. For ease of planting, lifting and other operations, soils of low clay and high sand contents are ideal, provided they are sufficiently deep, well drained, clear of perennial weeds, and free from stones and clods (which damage bulbs at lifting). Clearly not all these factors can be ideal, but as a general guide, good 'potato soils' are considered to be highly suitable for bulbs.

Commercial narcissus growing requires investment in specialized machinery and equipment; the crop is generally handled in bulk, and the quantities involved are often large, as lifted weights of over 40 tonnes ha^{-1} are not exceptional. Machines are available for planting, lifting, cleaning and grading, and it is necessary to have facilities for drying bulbs in bulk or in trays, insulated temperature controlled rooms for storing the harvested bulbs, tanks for hot-water treatment (h.w.t.) and an adequate supply of containers.

Planting

In the UK, narcissi are nearly always grown as a 2 year crop, in ridges, planted with a special bulb planting machine (planter) or a modified potato planter. This is for economic reasons; the improved yield and better disease control resulting from annual lifting (which is general practice in The Netherlands, and done by a few British growers) is achieved at the cost of the extra lifting and replanting. Ridges are 75 cm apart, with the bulbs occupying a 20 cm band, and 13 cm deep, measured from the top of the ridge to the base of the bulb. Prior to planting, fertilizer application after ploughing is recommended at the following rates (as kg ha^{-1}) in the case of soils of low nutritional status; N: 125, P_2O_5: 150, K_2O: 300 and Mg: 60–90. If following other well fertilized crops in the rotation, these levels are reduced proportionately. Potash application is a recognized requirement for good bulb growth. Because of current concern with nitrate pollution, it is advisable to apply P and K as base dressing, and to top-dress with N in winter just pre-emergence.

It is vital that all planting stock is given h.w.t. to control stem nematode and basal rot, and this operation accounts for 15% of the total cost of growing the crop (see Chapter 9). Bulbs are planted from late August to mid-October when soil conditions allow. The grades normally used for planting are 10–12 cm (circumference), smaller bulbs being unlikely to achieve saleable size by lifting time, 21 months later. Some very large bulbs (over 16 cm) are also included to guard against 'loss of stock vigour' which is feared might result from continued replanting of small bulbs, although there is no scientific evidence for this.

A consequence of adopting a 2 year growing cycle is that it is not possible to maintain the optimum planting density in both growing seasons; the recommendation is therefore a compromise, which is considered more profitable than growing two one-season crops with the extra lifting and replanting costs. In The Netherlands, where the crop is grown on a single season basis, planting densities used are about twice those used in the UK. The optimum planting density depends on the relative values of the bulbs of the cultivar being grown and land costs, as the husbandry costs are largely independent of planting density. For an expensive new cultivar, lower planting densities are clearly better, but for the standard, popular cultivars, current commercial densities for bulb production are 12.5–17.5 tonnes ha^{-1} depending on bulb size. Where a greater proportion of large bulbs is required, lower planting densities can be used. Because of the number of variables involved, the prediction of a planting density for a given situation is extremely difficult, despite the use of mathematical models. Additionally, the responses tend to be insensitive to planting density over a fairly broad set of values near the optimum and also dependent on seasonal differences. The range given above is a useful guide which is sufficiently accurate as a general recommendation.

To avoid any possibilities of cross-contamination, it is recommended that stock previously infested with nematodes should be planted separately from clean stock. For efficient handling, as well as to minimize flower and leaf damage by h.w.t., it is usual to plant the various types of narcissi in the order in which they are given h.w.t., i.e. poeticus, short-cupped, long-cupped and lastly, the trumpet cultivars. The mechanical planters normally employed plant two rows at a time, the bulbs being fed from a hopper usually into a troughed belt to deliver the bulbs into furrows, which are then covered by discs to leave $c.$ 13 cm of soil over the bulbs.

Husbandry

Aspects of husbandry which require attention during the growing season include weed control, as weeds can reduce bulb yield and interfere with lifting the bulbs. An investigation of effects of weeds on the growth and development of the narcissus crop in Scotland by Lawson and Wiseman (1978) showed that in the first growing season after planting, there was earlier foliage senescence and reduced bulb growth. The effect carried over into the second year as shown by fewer flowers, smaller bulbs and a lower bulb yield despite the crops being maintained free of weeds in their second year of growth. Weed competition was effective only after the crop flowered and persisted until normal senescence. If weeds were removed shortly after flowering, losses could be prevented, but later removal accelerated crop senescence. Overwintered weeds had the greatest effect because their growth paralleled that of the crop. Normal commercial practice is to use contact herbicides in autumn and winter followed just before leaf emergence by a residual herbicide. At the end of the first growing season mechanical and/or a contact herbicidal treatment is essential. Depending on soil type, it might be necessary to reform the ridges which often become eroded and less well-defined during the first season's growth.

Irrigation is said to be beneficial during April/May if and when a deficit of 25 mm has been reached, despite lack of consistent supporting experimental evidence for any benefit in the UK. Whilst the plants are in flower, removal of unhealthy individuals, especially those with virus symptoms, and of the wrong variety, a process termed 'roguing' is practised. Because of high labour costs, roguing is less consistently done than formerly, despite such devices as herbicide impregnated 'gloves' used to wipe unwanted plants, rather than physically digging them up with a roguing iron.

Experiments on deheading (removing the flowers, but leaving the scapes) have given widely different estimates of the effect on yield, compared with allowing the flowers to senesce naturally, from 20% in Cornwall to only 2% in The Netherlands, the difference probably being accounted for by leaf diseases in the wetter Cornish climate being encouraged by the decaying flowers. Nowadays, flowers are not 'deheaded' commercially, despite the increased bulb yield because this benefit does not repay the cost of flower removal, except in wet climates or seasons, when leaving the flowers increases the damage from *Botrytis* infection. In the Southwest, fungicidal spray programmes are effective in extending leaf longevity, and thereby increasing yield. Late sprays in the second year are avoided to prevent senescence being too delayed, and the start of lifting being hindered. Picking the flowers reduces bulb yield but the revenue gained from their sale greatly outweighs the value of the lost yield. It is important not to damage the foliage when picking flowers to reduce yields further because of loss of photosynthetic area and fungal infection of damaged leaves.

Lifting

To aid lifting, especially of bulbs required for early forcing or intended for export, it may be necessary to use a desiccant spray to kill off weeds and non-senescent foliage. Sulphuric acid is often used, but only by contractors because of its corrosive properties and the danger when mixing it with water. It is important not to use translocatable herbicides which can damage the bulbs and produce malformed foliage and flowers. Alternatively, mechanical flailing or shallow rotovation can be used to remove aerial growth, but to avoid damaging the bulbs this must not be too deep; it is recommended to leave 5 cm or so of leaf above the bulb tips.

Bulbs are lifted when at least 95% of the foliage has died down, and preferably at full leaf senescence, in mid June to late July. Special machines for lifting range from the simple elevator which lifts the bulbs, separates most of the soil then deposits the bulbs on the ground for hand gathering, to more complex machines (complete harvesters) designed for bulb use or modified from potato harvesters. It is important that the machines are carefully adjusted and operated so as to avoid the bruising and cutting damage to the bulbs which can become sites for later pest and disease activity.

Bulb Treatment

The bulbs are then transferred to a store, which may be a large bulk store or an open shed roofed to keep them dry and prevent sun scald or rain stain, both of which affect visible quality and saleability. Bulbs may be in nets, considered only as a short-term measure, be held in bulk bins (usually of half-tonne capacity, in which they can be treated from the harvester until graded) or be taken by trailer and be elevated after partial cleaning to a bulk store with air ducting for forced-air drying, as used for onions. On going into the store, bulbs are frequently given a fungicidal spray. Bulbs requiring special treatment such as preparation for forcing are cleaned and graded first, followed by the saleable stock. This leaves the planting stock for h.w.t. There are four main methods of handling narcissus bulbs, using nets, trays, bulk bins or loose, in bulk. The adoption of a system depends on the volume being handled, the other crops grown on the holding, labour availability, and the other equipment available, like tractors, trailers, elevators, fork-lift trucks, etc. Each system has its advantages and disadvantages.

Large quantities of bulbs are partly cleaned then dried in a bulk store or in bulk bins prior to a final cleaning and grading. Smaller quantities can be dealt with in one operation, using a vibrating riddle or a barrel cleaner, followed by passage over an inspection belt for sorting out remaining clods, debris and diseased and damaged bulbs, and manual separation of large bulb clusters. Next comes a grader to divide the bulbs into size categories (circumference) differing by 2 cm, such as 8–10, 10–12, 12–14, 14–16 and 16 + cm.

Bulb drying is a complex process. Initial drying is from the superficial parts of the bulb, but later water loss is dependent on the rate of migration of water from the inner scales. During the first phase the rate of water loss depends on the air flow and its temperature, and the recommended rate is $425 \, m^3 \, h^{-1} \, tonne^{-1}$, which will surface dry the bulbs in 53 h at 30°C. For the second phase, a lower flow rate of $170 \, m^3 \, h^{-1} \, tonne^{-1}$ is adequate with r.h. of 80–85%.

Between lifting and grading, the preferred temperature for bulb storage is about 18°C. This avoids the risk of flower damage caused by h.w.t. of bulbs stored below this temperature, and minimizes the risk of basal rot spread, which is greatest at 21–30°C. After grading, the storage treatments given depend on the grades and their future use.

Bulbs for sale are usually 14 cm or larger, and remain at 18°C in trays or nets, adequately ventilated to prevent premature rooting. Bulbs intended for forcing are usually 12–14 cm grade. For early forcing these receive 35°C for 5 days immediately after lifting, and are kept at 18°C until they reach Stage Pc, when the temperature is lowered to 9°C. Later-forced bulbs do not have the 35°C treatment or the 9°C, but remain at 17°C until planted. Planting stock, the large bulb clusters and the 8–12 cm offsets, are stored at 18°C to prevent serious subsequent h.w.t. damage, and are planted immediately after this treatment without drying.

Bulb yields for a two-season crop vary considerably with cultivar and season. Yields are frequently expressed as a percentage increase, i.e. 100 × (weight lifted − weight planted)/weight planted, and in these terms should give values of 100–180% for a good commercial cultivar, i.e. lifted weights of 25–50 tonnes ha^{-1}. To indicate the approximate bulb numbers involved, there are 38 000–56 000 8–10 cm bulbs per tonne, and 11 000–13 000 14–16 cm bulbs per tonne. The smaller figure of each pair refers to bulbs graded on a slotted pin grader, and the larger to a grader with round holes.

Bulbs are marketed according to customer requirements, in bulk and by weight to growers and wholesalers, in smaller units either bagged or loose for customer self-service to garden centres, flower shops and the like, whilst individual pre-packs are prepared for gardeners, usually with planting instructions and a coloured illustration of the cultivar. Whilst in commercial practice bulbs are sold by size grades, the older Dutch descriptions of bulb grades: double-nosed, single-nosed, round, offsets and chips, have now almost disappeared from commercial use.

Tazetta Narcissi

These narcissi, which are not generally winter hardy, are grown on the Isles of Scilly for their scented, early flowers. They have no cold requirement, so their culture differs markedly from other, hardy members of the genus. Bulbs are planted by August because rooting starts early, with leaf emergence in October. The foliage of plants in their second and third years appears in August and September, necessitating early application of herbicides. Planting densities of $15–20$ tonnes ha^{-1} are usual, not very different from those used for other narcissi. Bulbs are often left in the ground for several years, provided flower numbers are sufficiently high, but when necessary they are lifted in May/June. After lifting, early flowering is encouraged by storing the bulbs at $20–30°C$ for 4–6 weeks. Following this they should receive h.w.t. and be planted immediately afterwards. A treatment having the same effect as warm storage is the burning-over of the soil, originally using straw, but now using propane. The benefits are thought to be related to ethylene, as described in Chapter 5.

OTHER 'BULBS'

Tulip

Like most spring flowering bulbs other than narcissus, tulips are grown on a 1 year cycle. They are most successful in areas drier than those growing narcissi because there is then less damage from *Botrytis tulipae*, and planting and harvesting are also easier. Tulips also require well drained, lighter soils with a pH not below 6.5, a low clay content, and freedom from the stones and clods which can damage the delicate bulbs. In the UK they are usually grown in ridges, elsewhere beds are more common (Figs 7.3 and 7.4). General fertilizer recommendations are for 100 kg P_2O_5 and 200 kg K_2O ha^{-1}, modified in relation to soil analysis. Fertilizer is applied after ploughing and worked in prior to planting.

Tulips are planted later than narcissi because they benefit from a long

Fig. 7.3. Tulips in beds in The Netherlands. These have been deheaded.

period in a warm store, emergence is delayed and frost damage decreased, and, most importantly, there is less damage from *Fusarium* disease, which is favoured by warm soils. Bulbs are therefore planted, after a fungicidal dip, in the period October–early December, when soil conditions permit. Planting in the UK is in ridges, slightly narrower than narcissi at 71–75 cm apart, and less deep at 7.5–10 cm from the base of the bulb. Bulbs are usually in the size grades 7–10 cm (circumference) with a proportion of larger bulbs included to maintain quality. Recommended planting densities depend on cultivar and on bulb size, within the following numbers (as million bulbs ha^{-1}), based on, respectively, a vigorous cultivar producing large bulbs, and a more average cultivar: 8–9 cm, 1.06–0.94; 10–11 cm, 0.65–0.51; 12–14 cm, 0.55–0.33. Deheading after roguing is beneficial for disease avoidance and improved yield.

Herbicides are used to control weeds, and eliminate this competition from planting time until June. Whilst late season weed growth has no large effect on yield, it can interfere with lifting. A non-selective contact herbicide is usually applied after planting and a residual herbicide just before emergence. Whilst some growers irrigate their crops, there are dangers of increasing the incidence of fire disease, and, it is said, of uneven senescence, variability of forcing behaviour, and reduced flower number.

Tulip bulbs are lifted after complete senescence of the aerial parts; the

Fig. 7.4. Tulips in ridges in Lincolnshire being deheaded by machine. This operation requires manual back-up because not all flowers are the same height.

tunic is then partly or completely brown. For early forcing or for producing high quality tunics, lifting is earlier, before the tunic starts to turn brown. This introduces a risk of hard base or impeded rooting. Bulbs are lifted by an elevator digger which deposits the bulbs on the soil surface for gathering by hand, or by a complete harvester which takes the bulbs over an inspection belt to trays. At lifting, it is important that the bulbs are handled gently to avoid damage, so bulk handling methods are not generally used. Many growers wash the bulbs immediately after lifting to improve the appearance of those intended for the retail trade. The individual bulbs need to be separated from the clusters and the remains of the mother bulb, roots and old stem. This can be done by machine or by hand. The grades of bulbs vary somewhat with cultivar, but in general those for forcing are 11–12 and 12–13 cm, for the retail trade are 10–11 cm and above, whilst those for replanting are 7–8, 8–9, 9–10 and 13–14 cm.

Yields vary widely between cultivars and seasons, with percentage increases of between 60 and 150, and lifted bulb numbers of grade 10 cm and above of $125\,000$–$500\,000\,ha^{-1}$.

Storage temperatures for bulbs intended for forcing are discussed in Chapter 8. Bulbs for replanting are usually kept at 20°C and an r.h. of $c.\,75\%$. Research has shown that warmer temperatures result in later emergence, flowering and senescence in the field, and more flowering plants in borderline sized bulbs. Earlier recommendations based on dividing tulips into

Fig. 7.5. Field grown irises in western Washington, USA.

three response groups on the basis of their reaction to different temperature sequences have now been abandoned as being too variable and complex for general use. However, there are some possibilities for modifying field behaviour by storage treatments. For late-planted bulbs, there are benefits of improved rooting from reducing the storage temperature from 20 to 17°C from 1 November. Cultivars that tend to produce few offsets benefit from 6 weeks at 30°C immediately after lifting. A high-temperature treatment applied during storage to kill the flower bud, termed *blindstoken*, is sometimes used to increase bulb yields and the numbers of large bulbs by *c.* 20%, but the treatment is difficult to apply and there are dangers of yield loss by unintentional damage to the foliage. For one cultivar, the best pre-treatment temperature was 17°C, with 33°C being applied for 7 days at the end of September. It is not known whether other cultivars would respond better to other temperatures or times.

Bulbous Iris

The area of iris growing in the UK is only a fraction of that in The Netherlands. Other major producers are western Washington (USA) (Fig. 7.5), France and Israel. The production of iris bulbs for sale depends on the trade requirement for round bulbs of grade 8–9 cm and above, one of which is potentially produced by each non-flowering plant, provided the planted bulb is of an appropriate grade. The bulbs of flowering plants are flattened on the side next to the mother plant stem, called 'flats', have fewer protective outer scales and are consequently more easily damaged physically, are prone to dry in storage and often become infected by *Penicillium*. With care they can perform satisfactorily. Bulb production of cultivars with large bulbs, such as Ideal, depends on planting 5–6 and 6–7 cm bulbs. Smaller planting stock results in small rounds, with a lower flowering percentage.

Bulbs are planted in October/November, at densities of *c*. 1 million ha^{-1}, in ridges or beds. Depending on location, weather and planting date, leaf emergence can occur before the cold weather of winter, or be delayed until early spring. This makes weed control difficult, and usually involves applying a post-emergence herbicide, especially if a spring flush of weed growth is expected. The bulbs are lifted in the following July/August, just before complete senescence, and those intended for preparation and early forcing require rapid drying, combined with a heat treatment at 30°C with a good airflow. Bulbs smaller than 8 cm intended for replanting are kept cool and well ventilated to lower the flowering percentage in the field and thereby produce more flats. The optimum storage temperature for planting stock is 12–15°C, with some cultivar differences.

Lily

Very few lilies are grown as field crops for bulb production in the UK; most are imported, mainly from The Netherlands, which provided for the UK and the Channel Islands in 1989/90 over 37 million bulbs, of which about two-thirds were forced. Trials in Cornwall indicated that successful crops could be grown for flower production as well as for bulbs; trials in Lincolnshire look promising for bulb production.

Propagation of lilies is from seed of true-breeding species, from offsets and stem bulbils, or bulbils produced by scaling and micropropagation (see Chapter 6). A fertile well drained soil is recommended, with a pH of 6.0–6.5. Nutritional requirements are similar to those for narcissus. Bulbs can be planted in October/November or, following overwinter storage at −1°C, in March. The advantages of spring planting include the flexibility of planting date and avoidance of frost damage at emergence. The planting rates are 3–5 tonnes ha^{-1} depending on grade between 8 and 12 cm for 1 year crops, and the ridge system 76 cm apart is used in the UK to facilitate lifting. Bulbs are lifted in September, after flailing to remove the old stems. The bulbs are easily damaged and require careful handling. They are placed in trays and protected from drying too rapidly, until stored at 2°C. They are removed from store for cleaning and drying, which is mainly done by hand, then frozen at −2°C in polythene lined boxes of peat or vermiculite prior to storage at −1°C, whether required for forcing or field planting. On removal from the store, the bulbs are thawed at 15–20°C for several days, care being taken to avoid damaging the frozen bulbs. They sprout rapidly after thawing, and need planting immediately. The extended growing season means that weed competition could be serious, making effective herbicidal treatment essential. For further information see MAFF (1984a).

The 'mother blocking' system of maintaining superior, healthy true-to-type bulbs was developed for Easter lilies in the USA. A mother block comprises a grower's very best, individually selected plants, which are marked in the field, and inspected periodically during the season to discard any off-types, diseased or slow-growing individuals. At harvest these individually dug bulbs are kept separate and again subject to selection on the basis of bulb size. They are replanted separately the next season, and again carefully inspected and rogued. After 2 years, the plants that remain are grown together as a block, in isolation from other lily stocks, and referred to as a 'superior stock'. Similar procedures are used for other bulbs to produce 'elite stocks' or 'greenstocks', whose performance is substantially better than those of the original stock from which they were selected.

Anemone

Anemones are grown from corms (strictly tubers) or from seed. For seed production, the methods used for growing the plants are the same as for commercial flower production. Because anemone flowers are always sold as mixed colour bunches, it is necessary to grow families of the constituent colours separately until a final mass seeding prior to the sale of corms. In this way a good colour balance can be maintained. A spring flowering crop is used to produce seed for sowing in late summer. Corms of 2–3 cm grade are planted in early June in drills 2–5 cm deep at about $150\,000\,ha^{-1}$; the resulting plants start flowering in early June. Seed heads develop over 3–4 weeks, and soften to release the fluffy achenes. About 1500 seeds are produced per inflorescence with c. 50% viability. Seed yields of $60–160\,kg\,ha^{-1}$ are normal, depending on a range of factors. Seed has to be collected daily, dried and defluffed, either by hand or mechanically.

Corm production results from this seed, sown in frames at $5\,g\,m^{-2}$ in August, and grown overwinter under supplementary lighting until the foliage senesces in June/July. The small corms are recovered, dried and stored at 18–24°C for later planting in stone-free soil to aid corm recovery at lifting prior to sale. Further details on anemone growing are to be found in MAFF (1977).

Hippeastrum

In contrast to the narcissus, tulip and other spring bulbs, where production is outdoors in a temperate climate, the cultivation of hippeastrum is either in heated glasshouses as in The Netherlands or outdoors in warm countries like Israel or Swaziland. The starting material can be seed, or, more usually, offset bulbs or twin-scales (see Chapter 6). From offsets it takes 2 years to produce saleable bulbs, from seed germination 18–24 months, and from twin-scales 3 years. Offsets are planted in the greenhouse in November, are lifted 9 months later and replanted for a further two seasons. In February the bulbils resulting from twin-scaling in November are planted and allowed to grow until September. After lifting they are replanted in November for two standard 9 month growing cycles. About 30 bulbs m^{-2} is the standard planting density for 22 cm bulbs. After each lifting, which is done annually, planting stock is stored at 23°C for 7–10 days with good ventilation, before further storage at 13°C until replanting.

For field growing in the southern hemisphere, well drained beds are used. Offset bulbs are planted in July and grown until the following May. Prior to lifting, the leaves are removed. The lifted bulbs are washed, dipped in fungicide and dried before storage at 30°C for 10 days, then held at 7°C to prevent premature elongation of the flower scape.

Gladiolus

Gladiolus corm production starts with either cormels or corms. As flowers are produced in the field, the field crop grown from corms yields both flowers and corms, although harvesting procedures can be modified to favour one or the other. In American practice, cormels are graded into three sizes, below 0.6 cm, 0.6–1.0 cm and over 1.0 cm (the measurements being corm diameters, rather than circumference, as used for most 'bulbs'), and are produced from 'mother blocks' of plants carefully selected for disease freedom and grown in isolation on sterilized soil. Cormels are planted in single rows 10–13 cm wide and 60–70 cm apart in well drained soil at a depth of ca. 8 cm prior to levelling and compression. High planting densities are used (2–12 million ha^{-1}) and the resultant corms of 1.3 cm diameter and larger (perhaps 1.5 million ha^{-1} from large cormels), called planting stock, are grown on for a further year to give up to 430 000 flowering size corms (over 2.5 cm diameter).

Flower removal increases yield, but is usually delayed until the first floret opens to allow roguing. Harvesting the spikes for market, leaving four leaves, reduces corm yield by 30% compared with those with intact leaves, whether the inflorescence is removed or not. Removing the inflorescence as soon as it appears doubles the yield of cormels; corm yield is not significantly affected. In Israel, most of the growth of the corms, and all that of the cormels, occurs during a period of about 5 weeks starting when flowering ends. Corms lifted in the warm part of the year require 3–4 months at 2–4°C to break dormancy, or this may be achieved using ethylene chlorhydrin. In warmer parts, such as Florida and countries bordering the Mediterranean, gladioli are often grown as a winter crop, but where winter temperatures are below freezing, planting is delayed to avoid frost damage.

Tuberous Begonias

These are produced mainly in Belgium and California, and require cool night temperatures below 16°C for best growth. Tuberization is controlled by short days, on plants grown outdoors from seed. The tubers are lifted and cleaned and are on sale from about mid-January. For UK growers, the tubers are started off into growth in February/March at 10–16°C and moved to patio conditions or outdoors when there is no further risk of frost.

'BULB' STORAGE FOR REPLANTING

Storage temperatures for 'bulbs' intended for replanting are well established for the major 'bulbs', and successful for a range of minor bulbs as determined empirically.

Narcissus bulbs can be stored at about 18°C, which is a compromise between the requirements for a sufficiently high temperature to prevent h.w.t. damage and a sufficiently low one to reduce spread of *Fusarium* infection. Tulips are stored at 20°C, although 23°C is used to delay emergence and prevent frost damage. It is normal practice to lower these temperatures to 17°C for a month before planting. Hyacinths for planting are treated similarly, initially at 20°C (4 weeks) then 17°C. Advice on gladiolus corm storage is confusing. Some suggest 15–17°C after an initial warm period at 20–25°C, others suggest 4–6°C for large, cool-grown corms and 10°C for dormant ones lifted in warm weather. Cormels are given 10°C until a month before planting, then 20°C, according to Dutch recommendations, but American growers apparently use 2–4°C until planting. These differences probably reflect the widely different climates in which gladioli are grown in several parts of the world, and the empirical, but successful, growing systems developed in each. For iris, several temperatures have been suggested: 20°C for large or mixed sized bulbs, but 12–15°C (or even 10°C) for small bulbs to prevent them flowering. Freesias require 3–4 months at 30°C. In contrast, lily bulbs are stored at −1°C in moist peat. Requirements of minor bulbs are shown in Table 7.1.

WEED CONTROL

Two essential considerations in choosing a herbicide are its safety at the

Table 7.1. Recommended storage temperatures for minor bulbs. Most are based on empirical experience rather than on experimental work. It must be remembered that there is a danger that small bulbs lose considerably more water than larger bulbs do because of their larger surface : volume ratio. This is particularly true if they are stored with other bulbs which require ventilation and dry conditions. An asterisk is used when the recommendation is for storage in some packing material such as peat or wood shavings to restrict water loss.

Acidanthera	20°C
Allium	20–23°C
Allium giganteum	25–27°C
Anemone blanda	17–20°C
Anemone coronaria	13°C
Arum cornutum	2–5°C
Brodiaea	20–23°C
Camassia	20°C
Chionodoxa	20°C
Colchicum	17–20°C
Crocus	20°C
Dahlia	5–9°C
Eranthis*	5°C
Eremurus	5–9°C
Erythronium*	5–9°C
Fritillaria imperialis	25°C
Fritillaria meleagris*	9°C
Galanthus*	17°C
Galtonia	17–20°C
Hymenocallis	20°C
Iris danfordiae	23–25°C
Iris reticulata	23–25°C
Ixia	23°C
Muscari	23–25°C
Oxalis adenophylla	17–20°C
Ranunculus	13°C
Scilla	23°C
Sparaxis	25°C
Sprekelia	20°C
Tigridia	2.5°C
Tritonia	25°C
Zantedeschia	13°C

relevant stage of the crop's growth and its efficacy against the weeds that are present at application or are expected to germinate subsequently. For contact herbicides there is often a limited 'window' in the crop's development, outside which severe damage can result. Residual herbicides have less restricted application periods, but for autumn planted 'bulbs' there is a long period of gradually deteriorating weather and wet soil conditions which can make application difficult. These problems can to some extent be overcome by spraying later, when soils are frozen and capable of supporting the weight of a tractor. Application treatments can be classified as follows.

1. Pre-planting. Aimed at controlling perennial weeds, application must be sufficiently early to allow a safe period before planting to avoid crop damage.
2. Contact pre-emergence. For the control of weeds which germinate and grow before crop emergence.
3. Residual pre-emergence. Best applied to weed-free soil, it can be combined with 2 above.
4. Residual post-emergence. As an alternative to 3 applicable when the crop is at an early stage of growth, about 10 cm leaf emergence of narcissus and iris, and before tulip leaves have unfurled.

Additionally, a limited number of herbicides can be used just before and after flowering, if at low concentration, where other methods used earlier in the season have for some reason not been sufficiently successful. Herbicides are also used to control weeds in late summer after the first growing season of a 2 year narcissus crop. Application is made after the aerial parts have died down completely, and mechanically flailed to remove debris. A contact spray is used for annual weeds and translocated herbicides for controlling perennial weeds.

Inevitably, some bulbs are left in the ground after lifting and their persistence lessens the efficiency of crop rotation. Such 'groundkeepers' can be eliminated using a non-selective, foliar acting contact herbicide.

In general, residual herbicides should be sprayed onto moist soil, so that they reach weed seeds germinating below the surface. For most 'bulb' crops, conditions at the appropriate times are favourable, except for gladioli in some seasons. Soil type affects herbicide effectiveness; for instance on highly organic peat soils, many commonly used herbicides are less persistent than on the silts and loams where most of the testing was done. This is because the absorptive capacity of a soil for herbicides increases with its organic matter content. Manufacturers' label recommendations usually specify dosage in relation to soil type, and also provide information on resistant weeds.

MECHANIZATION

An important feature contributing to successful and economically efficient bulb growing in the UK is that most operations are highly mechanized. The quantities of 'bulbs' handled can be large and there is urgency for many of the standard operations because of weather. It is necessary to separate lifted stock after cleaning into categories for sale, for replanting and for forcing, and there is a requirement for timely h.w.t. and to start temperature treatments early. A recent study of gladiolus growing in Sicily showed that mechanical planting

Fig. 7.6. A twin-row bulb planting machine. The large hopper feeds the bulbs into the bases of the two furrows formed by the two coulters, and the two paired discs (one is seen in front of the wheel) form ridges over the two rows of bulbs.

was 7.5 times and mechanical harvesting 3.3 times faster than manual operations. For more information on 'bulb' mechanization in the UK, see Balls (1985, 1986).

Planting and Lifting

Mechanical planters are now universally used; some are specially designed for use with 'bulbs', others are modified potato planters (Fig. 7.6). Planting a large area has to be a well coordinated operation to deliver bulbs to the planter, especially as current trends towards higher planting densities require more frequent refilling of the planting machine hopper. Similarly, lifting is a large-scale operation, aided by preparatory work to pulverize dried plant material and weeds on the soil surface and to break up clods and remove the ridge tops (Fig. 7.7). Because narcissus bulbs are better able to stand up to rougher handling than other bulbs, they are generally harvested in bulk using a complete harvester (Fig. 7.8). Growers of other 'bulbs' and small-scale growers of narcissi use elevator diggers to lift single ridges, or two at a time, and deposit the 'bulbs', after agitation to remove much of the adhering soil and clods, on the soil surface. They are then picked up by hand into trays and removed from the field. Small 'bulbs', such as anemones, require specialized machinery to separate the 'bulbs' efficiently from the soil without losing too many (Fig. 7.9).

Fig. 7.7. (a) A Grimme bulb lifter in a narcissus field. (b) This machine lifts two rows of bulbs, carries them up an elevator with rubber-covered horizontal bars, then deposits the bulbs, shaken free of much adhering soil and debris, back on the soil surface in a single windrow to ease subsequent collection. (Photographs courtesy of Richard Pearson Ltd, Boston, Lincs., agents for Grimme.)

Fig. 7.8. An elevator–lifter for bulbs. The bulbs are lifted and conveyed along the overhead elevator and deposited into a trailer or bulk bin running alongside. Before reaching the container, much of the soil and debris is shaken free, and deposited on the soil behind the machine.

Fig. 7.9. A lifting machine designed for anemones, but suitable for other small bulbs. Note the fine mesh screen for separating the 'bulbs' from the sandy soil.

Fig. 7.10. Standard bulb trays containing iris bulbs. Note the gaps between the trays allowing good air movement, the strong corners allowing the trays to be nested and the carrying handles.

Handling

Large 'bulbs' are generally handled in trays or in bulk boxes. Trays are of wood or plastic, designed to stack vertically leaving a good air space between them (Fig. 7.10). They are of two sizes, 75×45 cm and 60×45 cm, the former being the standard potato chitting tray. As the larger tray has shallower sides, there is not much difference in the weights of bulbs accommodated in each, and a rule-of-thumb value is 80 trays per tonne of 'bulbs', as the bulk density of loose bulbs is $c. 600$ kg m^{-3}. It is normal to allow one-third extra space in the store for access, so the total requirement is $c. 6$ m^3 tonne^{-1}. Smaller 'bulbs' are frequently handled and stored in 30×45 cm trays. Trays are normally moved on pallet bases, four to a layer, and can be stacked up to eight high.

Bulk containers are used mainly for narcissus and gladiolus 'bulbs'. They are of wood and have a capacity of 0.85 m^3 (0.5 tonne of 'bulbs'), and a base of 1.2×1.0 m to fit standard BS pallet bases. They are used for drying and subsequent storage and have slatted sides and bases. Special drying boxes have solid sides and special bases so that these form an airtight plenum duct when assembled at a drying wall. Air is then forced through the bulbs. Bulk boxes are handled by fork-lift and pallet trucks, and require a means of emptying them, using a static tipping cage or a fork-lift.

Cleaning

The amount of soil harvested with 'bulbs' depends on soil type, its moisture content at harvesting, and the method of lifting. Some soil will be lost during drying, and the greatest problem is that of hard clods of bulb size or larger. There are several methods of breaking these but all result in some bulb damage. Rotary barrel cleaners, agitated chain conveyors and cleaners with star shaped spools are all used, and their designs are being improved to lessen damage to the 'bulbs'. Much of the tulip and lily production, as well as that of the smaller 'bulbs', is cleaned by hand, by casual labour, in the yard or shed.

Some cleaning occurs after drying (see below), when the 'bulbs' are on the grader, where they pass first over a riddle which removes soil and debris (scales, stalks, dried foliage). At this point a dust extractor contributes to operator comfort and crop hygiene. A system of brushes is often incorporated early in the grading line to clean tulip bulbs and split up the clusters. The 'bulbs' usually pass over a roller inspection table, which turns them so that all sides are visible, prior to grading; any damaged or diseased bulbs can then be removed.

Drying

Depending on the weather, species and the lifting method, some 'bulbs' are allowed to dry for a few days on the soil surface (windrowing). This is common practice for narcissus in the Southwest, but not in Lincolnshire because of the danger of sun scorch in sunny weather. Tulips are never dried in this way because of the high risk of damage. Open sided, covered buildings are safer, and also provide protection from rain. Many 'bulbs' are dried in simple, inexpensive, insulated, mechanically ventilated stores which can also serve for subsequent storage. Bulk quantities of narcissus bulbs can be dried successfully in bulk boxes linked to a force-ventilated drying wall (often called a letterbox wall because of the shape of the outlets against which the boxes are stacked for forcing air through the bulbs), or in bulk, loose, up to a 3 m depth on a floor provided with ducted forced-air ventilation. 'Bulb' structure determines drying methods; tulips have only a single tunic so that before the top layers of bulbs are dry, those near the inlet will have been over-dried, and will have shed their tunics. For this reason, tulips are normally dried in shallow layers at ambient temperatures and 65–70% r.h. 'Bulb' drying is a complex and slow process, requiring careful temperature control to avoid the spread of fungal diseases such as the basal rot of narcissus, as discussed in Chapter 9. The best system of drying for an individual grower depends on many factors, such as the quantities and range of species and cultivars grown, the marketing outlet being supplied, other ancillary equipment available, etc.

Fig. 7.11. A bulb grader. Bulbs are loaded into the hopper on the left, and move by agitation over a series of grids (riddles), through which the bulbs fall into canvas funnels into trays or other containers. The riddles are slotted (bars) or with round holes, shown propped up behind the bulb tray, and appropriate sizes are available for most bulbs. In the grader there are four riddles, which can give five grades, including those too large for the largest riddle which pass over the end of the machine.

To develop such a system, a horticultural engineer should be consulted at an early stage to take into account all the relevant factors.

Grading

'Bulbs' are usually graded into a minimum of four sizes, irrespective of their shape (round or flat sided). Most present-day graders use a series of interchangeable riddles (screens) held in horizontal frames (Fig. 7.11). The 'bulbs' are progressed along the frames, starting with the smallest, by a jumping action, until each encounters a hole through which it falls, into a tray or net. Provision is also made for 'bulbs' larger than the largest screen. The screens have round holes for general use, but for narcissi slots or parallel bars are usual. A flat sided 'bulb' which would not fall through a round hole would do so through a slot when aligned longitudinally. For a given screen size, therefore, 'bulbs' graded on a slot are larger than those graded on a round hole. Ancillary facilities are often used to count tulip bulbs or weigh narcissus bulbs, as part of the grading process.

Buildings

For a given farm, with a known area of 'bulb' growing, it is necessary to provide adequate storage facilities with sufficient flexibility to be able to cope with all the temperatures required for all the quantities being stored, without having expensive capacity which is only rarely required, and idle for much of the time. The other crops being grown also have a bearing on efficient use of storage space, so that they can be used during the period when not in use, or only partly used, for 'bulbs', i.e. late winter to mid summer. Whilst stores are becoming increasingly more sophisticated, there is still widespread use of 'open' stores, using natural ventilation for storing 'bulbs' prior to replanting.

Although there is a range of requirements for 'bulb' storage, they can be divided into the two broad categories 'warm' and 'cool'. Examples of the former are:

1. narcissus: 35°C for advancing flowering, 30°C for pre-h.w.t., 17°C for pre-planting;
2. tulip: 35°C for preparation for early forcing, 20°C for pre-planting storage;
3. iris: 35°C and 40°C to improve flowering, 25°C for retarded bulbs.

Lower temperatures are required for preparing bulbs for forcing, 9°C being used for narcissus and tulip, 5°C is also used for tulip forcing and temperatures near freezing are required for storing lily bulbs and 'ice' tulips. Temperatures c. 1.5°C are necessary for short-term storage of cut flowers prior to marketing.

This list is not exhaustive, but indicates the range of temperatures required. There is also a requirement for r.h.; in general higher storage temperatures require associated high humidities, to prevent excessive drying of the 'bulbs'. There is a shortage of critical information, but 90–95% r.h. is often recommended. In contrast, for the lower temperatures, low humidities are necessary to prevent rooting prior to planting. Humidity can, to some extent, be increased locally by enclosing a part of the store with plastic sheeting, or lining and covering boxes with plastic, e.g. for cut flowers, which require low temperature and high humidity.

Ventilation is another factor in the design of storage buildings, necessary to prevent the accumulation of ethylene from *Fusarium*-infected bulbs causing bud necrosis (see Chapter 9). As many 'bulbs' are sensitive to ethylene concentrations as low as 0.1 v.p.m., if 10% of 'bulbs' are infected in a store at 20°C, an air exchange rate of 10–15 times an hour is required. Air circulation is also necessary to maintain even temperature distribution within the store.

Buildings require good all-weather access, room to manoeuvre vehicles, and should be sited conveniently close to other operational areas whilst avoiding as far as possible the contamination of clean, healthy 'bulbs' with wind-blown debris from just-lifted stocks. Effects of noise and dust must be considered in relation to dwellings and staff. Floors of buildings must be sufficiently strong to take the weight of vehicles and their loads, and have suitable slopes and drainage to facilitate washing down. Doors must be large enough for access by lorries, elevators and other equipment. Ventilation is necessary, but birds and vermin must be excluded. A three-phase power supply is usually required, and good lighting (both natural and artificial). An important consideration is to allow for the possibility of later extension of the building.

If narcissi are grown, an h.w.t. facility is essential (Fig. 7.12). This must be separate from other buildings, and should incorporate a one-way flow system. Untreated bulbs enter at the 'dirty' end, pass through the room with the h.w.t. tanks, and leave through the 'clean' end. Sliding doors or flexible curtains each side of the central room will help reduce the spread of infection. If the tanks are loaded by hoist, then an overhead track is required, lighting and good ventilation are necessary, together with provision for cleaning down, washing the walls and floor and for disposing of debris. Separate vehicles should be used to deliver the 'dirty' bulbs and to pick up the 'clean' ones.

Specialized buildings for the treatment of 'bulbs' require sophisticated equipment to control temperature, humidity and ventilation. Insulation is important to reduce solar heat gain in summer and heat losses in winter, and for the same reason, these stores are often windowless and artificially lit. Because of high humidity inside these stores, the inner walls are usually specialized panels to resist moisture entry, and electrical fittings must be waterproof.

Because many 'bulb' crops can be forced in low light, the flower being produced at the expense of the food materials in the storage organ, insulated buildings have been used for this purpose, and sometimes the same building is used for storing the bulbs and then forcing them. A major advantage is the considerable saving of between 60 and 85% of heating energy during this winter/early spring period because the building is much better insulated than in the normal heated glasshouse situation. Some light is necessary to allow normal cultural operations and finally to green the plants (otherwise the stems and leaves tend to be abnormally yellow and unattractive to the consumer), and this can be achieved by having 15% of the cladding light transmitting, or by providing artificial light in a totally opaque building. Whilst such measures are generally successful for narcissus and tulip, most

Fig. 7.12. A top loading h.w.t. tank for narcissus. The cage, here shown empty and suspended from the overhead hoist, is filled with bulbs in trays or nets, and is lowered into the water preheated to near the temperature required. The hinged insulated lid (white, foreground) is then closed and the water heated to the exact temperature by thermostatically controlled heaters in the bottom of the tank. The water is circulated by the pump on the left and debris is caught up on filters. Behind the tank is a smaller, pre-soak tank.

forcing of these crops is still done in glasshouses, although these are usually plastic lined to improve their insulation compared with single glazed structures.

MARKETING

In the UK, most bulbs are sold by private negotiation, although the specialized auctions, such as that in Spalding, are used by some growers. Most bulb sales are in bulk to forcers and to local authorities' parks departments who use large numbers of 'bulbs' annually. There is an increasing trend to sell in smaller quantities such as 20 kg nets from which customers in garden centres select their own bulbs into standard bags at a fixed price. Many of the larger firms are producing small prepacks each containing a few bulbs with planting instructions and an attractive colour illustration; these are retailed through many outlets, often attracting impulse buying at petrol stations and supermarkets.

FLOWER PRODUCTION

The end-point of growing 'bulb' plants is the flower, although many commercial growers have subsidiary, intermediate objectives like producing 'bulbs', seeds or other propagules for sale to others who then grow them on to produce flowers. A major concern of the flower grower is the timing of flowering, so that flowers are regularly and predictably available out of season as well as at the natural flowering time. 'Bulb' plants are ideal for this purpose because of their partial or complete independence from current photosynthesis for the development and growth to anthesis of the flower, the final saleable product.

Commercial flower production may be achieved using several methods; the bulbs, corms and tubers can be grown in the open, under protection of glass or plastic, with or without heating, for cut flowers or pot-plant sales. Flowering date may be 'natural season', i.e. controlled by ambient temperature, or it may be out of season, i.e. forced, by special treatment of the 'bulbs' prior to and/or after planting. For some species, planting material can be kept in store for sufficiently long to allow sequential planting and year-round flowering, as with bulbous iris and some lilies. In commercial conditions, it is common practice to force a succession of crops ('rounds') into flower; this ensures efficient use of glasshouse facilities and continuity of labour use. To achieve a long forcing season, it may be necessary to employ a series of different storage conditions for the plants to be housed at different times for early, mid-season and late forcing, and the number of rounds in a UK forcing season can be as high as nine, starting with narcissi in early November and ending with tulips in April.

For this chapter, the starting point is taken as lifting time, because for very early forcing, storage treatments of the 'bulbs' must start then, the grower dividing his lifted stock into those intended for replanting or outdoor flowering from those destined for forcing, as soon as possible, compatible with cleaning, drying and grading.

Some 'natural' methods for producing early flowers of tulip and hyacinth were in common use in the 18th century, and involved the transfer of boxed bulbs from outdoor cold into protection before the end of winter. The development of methods of forcing cold-requiring, spring flowering bulb plants stems from the acute observations of a Dutch grower, Nicolaas Dames, in about 1907. By lifting hyacinths early and storing the bulbs under artificial conditions – the earliest investigations leading to 'prepared' hyacinths – he extended their flowering period and allowed flowering by Christmas. Some earliness of flowering is readily achieved by bringing into warm conditions 'bulbs' planted in pots or other portable containers after their cold requirement has been fully satisfied by outdoor conditions in winter. This procedure is still employed as a cheap means of late forcing, using plunge beds where the boxed bulbs are buried under ashes or straw until late in the winter. More sophisticated techniques are required to produce earlier flowers, and for the accurate timing of successions of flowering dates, involving the use of controlled temperature facilities. These include: i) speeding the development of the shoot within the bulb by early lifting, rapid cleaning and grading, and storage at high temperature; ii) identifying the stage of completion of flower initiation, transferring the bulbs to low temperature to start the cold treatment; iii) planting the bulbs at the most appropriate time; iv) accurately controlling temperature to achieve the most rapid satisfaction of the cold requirement and identifying this stage; and finally v) transferring to the glasshouse at the optimum temperature to achieve anthesis at the earliest possible date. By these means, narcissi can be in flower by late October and tulips well before Christmas. Whilst this is a basic pattern, there are several major alternative routes to early forcing, involving, for instance, storage of 'bulbs' at near freezing temperatures for long periods, and withdrawing them when required, the provision of the whole of the low-temperature treatment to the dry 'bulb' before planting, the use of complex sequences of low temperatures, forcing under artificial lights in insulated buildings, and storing forced plants at low temperatures just prior to anthesis as a means of marketing on specific dates.

Because several plant processes are involved at any stage (e.g. after planting, leaf expansion, flower growth and rooting are all in progress) each with its own optimum temperature, successful temperature sequences are a compromise between the optima, often arrived at empirically. For grower use, forcing schedules are available giving data for individual cultivars grown for flowering at different dates, as pot plants or cut flowers using different forcing methods. The most comprehensive of these is the Holland Bulb Forcer's Guide (De Hertogh, 1989), but see also the various MAFF publications on flowering (e.g. MAFF, 1976; MAFF, 1981). Detailed information on post-harvest storage of flowers is available in Nowak and Rudnicki (1990).

In all these commercial operations, the costs of production are important, and must be weighed against the higher prices achieved from early flowering, and the lower prices resulting from possible quality losses caused by some treatments. The organization of forcing schedules is difficult but important, and there is considerable investment required in large storage buildings with accurate temperature control, containers for handling large quantities of 'bulbs' and means such as fork-lift trucks and roller conveyors for rapidly transferring quantities of heavy, filled, forcing trays into and out of glass-houses.

The more sophisticated the control of temperature, the more reliable is the success of flowering, and the timing of flowering can also be controlled reliably. Whilst the cold requirement of a spring flowering bulb cultivar varies very little between seasons, other factors do vary. The weather during the growing season, especially the date of leaf senescence, determines when flower differentiation is completed within bulbs of narcissus and tulip. This point in development generally indicates the earliest date for starting low-temperature treatment, especially for tulip. It can vary by several weeks between seasons, necessitating bulb dissections in summer to ascertain the progress of floral differentiation. When a forcer buys pre-cooled bulbs from a supplier, it is essential to find out when the low-temperature treatment was started, and hence how long a period of cold the bulbs have received.

A considerably cheaper and much less accurate method of providing the low-temperature treatment system, which relies on ambient temperature, is the standing ground, which requires more grower experience and expertise to produce good results, consistently. Standing ground temperatures can vary considerably from the required 9°C, especially early on, in a warm autumn. For tulips it is suggested that this shortfall in accumulated cold can be compensated by delaying the housing of the bulbs by 1 day for each degree-week above 9°C. For narcissus each week above 9°C requires an extension of 3 days. Such judgements, and the need to provide a little extra duration of cold 'to be on the safe side' or in lieu of actually measuring temperatures, are not necessary with the accurate temperature control provided by well run controlled temperature stores.

NARCISSUS FORCING

Most of UK forcing is of narcissus, mainly cultivars of Divisions 1 and 2 of the Royal Horticultural Society (the Trumpet and Long-cupped Narcissi of Garden Origin) and mainly yellow in colour, with smaller quantities of cultivars with double flowers and with more strongly coloured trumpets. A list of the commonly forced UK and Netherlands cultivars was given in Chapter 3. The majority of narcissus bulbs forced in the UK are for cut-flower production, with a small proportion marketed as pot plants. In other countries this is not so, and in the USA, the opposite holds. There is increasing interest in pot-narcissus production in the UK and The Netherlands. The account below deals first with narcissi with a cold requirement, but there are cultivars, based on *Narcissus tazetta*, which can be flowered with no requirement for a long cold period, by direct planting in a glasshouse, or in the home. This group will be described separately.

Pre-planting Treatment

For earliest forcing, it is essential that the bulbs receive a high-temperature treatment starting within 3 days of early lifting, followed by the sequence of an intermediate temperature to allow completion of flower differentiation (Stage Pc), planting, cool storage to satisfy the low-temperature requirement, and finally transfer to appropriate glasshouse temperatures. A typical time-table for very early flowering of UK-grown bulbs is as follows.

> Lift on 8 June, warm store at 35°C for 5 days, 17°C until 7 August, 6 weeks at 9°C, plant on 18 September, return to 9°C for a further 6 weeks, move to glasshouse at 16°C on 30 October, full flower on 20 November.

The physiological effects of the post-lifting, warm storage period are not understood, although its effects are readily apparent as 12–14 days earlier anthesis of forced flowers of the commonly forced cultivars. Neither the actual temperature nor its duration appear to be critical; temperatures of 30, 34 and 35°C have all been used successfully for durations of 2, 5 and 7 days. The procedures adopted by individual growers depend on the temperature facilities available and on durations that fit in with workload; a 7 day period avoids weekend work. The treatment has no effect on the date of completion of flower differentiation, and can be regarded as equivalent to the exposure to high temperatures received by the bulbs in summer in the soil of their native environment. It has been noticed that an early 'easy' forcing season follows above-normal soil temperatures in the previous growing season, especially shortly before lifting.

Whilst it is possible to produce even earlier flowers by fine tuning the timetable above (at least for some cultivars), by starting cool storage before Stage Pc is reached, by shortening the period at 17°C, and by using a temperature below 9°C for part of the low-temperature period, it is at the expense of increased risk of flower damage such as poorly elongated scapes, moribund flower buds or of producing flowers of low quality because of short stems and small flowers (Hanks and Rees, 1984).

For mid-season flowering a lifting date of mid-July is satisfactory, followed by storage of the bulbs at 17°C until Stage Pc, followed by 6 weeks at 9°C, planting and returning the boxes to 9°C or keeping them on an outdoor standing ground until housed. For late season flowering, bulbs can be lifted as late as the end of July, provided that they have not started to re-root, and kept at ambient temperature until planted (i.e. they are not stored at 9°C before planting). They are housed from a 9°C store when required, once the low-temperature requirement is satisfied.

Planting

Bulbs of grades 14–16 and 12–14 cm are generally used for forcing, with smaller bulbs (10–12 cm) of some cultivars for the latest forcing rounds. Mean weights of bulbs of these grades, as kg per 1000 bulbs (and g bulb^{-1}) are 82, 52 and 39 respectively. The bulbs are planted in forcing trays of wood or plastic with corner posts to allow them to be stacked whilst leaving space for shoot growth. The growing medium is either peat or soil, taking care for pest and disease reasons to avoid soil previously used to grow bulbs, and not using glasshouse soil of high conductivity because of the risk of its residual salt content damaging the roots. Boxing soil needs to have good drainage and high air porosity; a sandy loam is best. The bulbs are usually arranged as thickly as possible in a 6–8 cm layer and covered with a layer of sand to stabilize the bulbs and prevent them from ejecting themselves when the elongating roots meet the base of the box. Peat needs to be thoroughly wetted before use, and it is normal to cover the planted bulbs with a further layer of peat. Peat has the advantage of lightness, which is appreciated by those handling planted boxes, but it is more expensive. If peat is re-used for forcing, it should be sterilized by steam or chemically.

After planting and thorough watering, the boxes can be returned to the controlled temperature cold store ('double cooled') (Fig. 8.1) or kept outdoors at ambient temperature. The former is used for the earlier forcing rounds, the latter is sufficiently precise for the later ones. In the cold store, the boxes must be kept well watered to encourage rooting, and this maintains high humidity in the store. Outdoors, there are two commonly used systems, the traditional standing ground (Fig. 8.2) where the boxes are laid on level, free-draining ground and covered with a layer of straw 20–30 cm deep kept moist, usually with overhead spray lines to encourage evaporative cooling (if ambient temperatures are above 9°C soon after planting), and to promote rapid rooting. Keeping the boxes moist helps prevent them freezing, and possibly damaging the plants in very cold spells. Good contact between the underlying soil and the boxes, and the use of plastic covers in hard weather also helps.

The second system uses stacks of up to eight boxes high on pallets assembled in correct order for eventual removal to the glasshouse. Efficient watering to ensure rapid and uniform rooting is essential, but can be difficult to achieve, despite the use of a spray line sited over the top layer of boxes. The stack is usually surrounded by straw bales, with a layer of loose straw on top. Experience shows that temperatures inside boxes in a stack are slighly lower than those on a standing ground; this explains why their shoots are shorter when ready to house. With both systems some precautions are necessary. The straw used must not have been treated with herbicide, or contain residues which could damage the emergent shoots. Protection from rodents is essential.

Fig. 8.1. Narcissus bulbs receiving low-temperature treatment in cold store (also called a rooting room). Note the pallet to aid filling and emptying the store with a fork-lift truck. Stacks can be up to three pallet loads high (each eight boxes high) in the store.

Readiness to House

For the early forced bulbs treated at controlled temperatures, the stage of readiness to house is determined by the total duration of cold that the plants have received. The requirement is for 12–15 weeks at 9°C, depending on cultivar. The longer the cold duration, the shorter the time in the glasshouse to flowering; this is important because of the high costs of glasshouse heating, but the asymptotic shape of the curve relating time in-house to duration of cold received makes the total time from start of cold to anthesis increase after about 12 weeks' cold, so the date of flowering becomes progressively later (See Fig. 5.4).

For bulbs treated on a standing ground or in stacks, readiness to house can be assessed from the position of the flower bud in the neck of the bulb, determined by gently squeezing this region or by cutting open a few bulbs longitudinally. Growth of the scape to this stage is related to two opposing effects of temperature: satisfying the cold requirement and controlling extension growth of the scape, with the mean optimum for the two processes at about 9°C. If the temperature is much below this, then the scape could be short, although the cold requirement could have been well satisfied, so that growth will be rapid following transfer to a higher temperature. On the other

Fig. 8.2. Above: a standing ground for providing bulbs with their cold treatment after planting, in boxes, buried under straw. Each row comprises a double row of boxes. Below: a single box exposed to show the bulb-thick planting density used for narcissus bulbs.

hand, warmer temperatures will allow an appreciable extension which could be misleading in terms of subsequent performance if the cold requirement has been poorly satisfied. The position of the flower bud is therefore not a good indication of readiness to house. An alternative method is to record the temperature daily in a position representative of the standing ground or stack and use accumulated day-degrees below 30°C to assess the extent of satisfaction of the cold requirement, with *c.* 2000 units being needed for flowering in 30 days at 16°C and *c.* 2250 for flowering in 21 days.

Fig. 8.3. Boxes of narcissi in a 'Venlo' type greenhouse at the flower picking stage. Note the high-level heating pipes which leave the floor clear, the narrow paths between the boxes and the width of the area occupied by the plastic boxes which allows the pickers to reach to the centre. Approximately 70% of the floor space is used for cropping. The vertical walls of the glasshouse are polythene lined to save heat loss, but not the roof as this modern house has a thermal screen.

The Glasshouse Phase

Glasshouses are used for bulb forcing because they can also be used for other crops (such as tomatoes) outside the bulb forcing season. The light requirement for narcissus forcing is small, only sufficient to produce natural greening of leaves and scape. The glasshouses are therefore frequently lined with plastic film to help conserve heat, this more than cancelling the lost benefits of raised temperature from solar gain. The boxes are placed on the floor or on benches, with sufficient path space to allow for picking the flowers (Fig. 8.3). In some modern glasshouses, roller benches are used allowing spaces for picking by moving benches as required, and giving greater utilization of the glasshouse area. When fixed benches are used, a second batch of planted forcing trays is often placed under the benches to start into growth; these are raised onto the benches when the previous round is cleared from the house. Bulb planting densities used depend on size, typical values for the three grades 10–12,

12–14 and 14–16 cm being, respectively 28, 25 and 13.5 thousand bulbs per 100 m^2 including paths, or in weight terms, over a tonne per 100 m^2.

Mobile glasshouses are sometimes used, the house being moved over the standing ground, at the appropriate time, and then moved away when cropping is finished, when the bulb boxes are uncovered and left until the time of lifting bulbs in the field for recovering the bulbs.

The traditional glasshouse temperature for forcing narcissi is 16°C, but this can be varied slightly to hasten or delay flowering to meet special marketing dates, weekend demand, etc. Flower quality is improved, especially early in the season, by using slightly lower temperatures, especially when the boxes have just been housed. Care must be taken with watering during the glasshouse phase, and the atmospheric humidity must not be too low, as this tends to dry out the spathe, restricting flower opening, and can damage the perianth parts and produce brown tips to the leaves.

Harvesting and Marketing

The best time for picking flowers is when they reach the 'goose-neck' stage, when the flower has bent over slightly, the spathe has split, showing colour, but before the perianth parts separate. Picking earlier, at the 'pencil' stage, is often necessary for flowers intended for export, but results in smaller flowers which do not last as long. Flowers are picked by hand, the scapes being snapped off at the tip of the bulb, and the bunches are made up at picking, with the heads level and the stems held by two narrow rubber bands. Some leaves are usually picked with the flowers and are included in each bunch. Further attention to presentation, such as stem trimming, is done when the bunches are boxed for despatch to market, usually in pre-cooled boxes of 30 bunches. During picking, especially later in the season, the unpicked, slightly later, flower stems tend to flop, producing curved stems and problems for the pickers. Support with string around canes in the corners of each box elimina- tes the problem. Ethephon has also been used successfully as a soil drench on many cultivars grown in soil to prevent this excessive elongation of leaves and scape, as well as for producing more attractive pot plants.

Cut flowers can be stored before despatch to market without affecting subsequent vase life, either upright in water or dry, wrapped in plastic film, flat in trays or in open market boxes. Recommended storage durations depend on temperature: 10 days at 0–1°C, 8 at 3–4°C, or 1–2 at 10°C. For transport, the dry flowers, packed in moisture retentive boxes, should be at 1°C. Some cultivars develop curved stems during storage if kept other than upright, or during transport to market.

Flower yields depend on cultivar, season, bulb grades used and the dates of forcing. Typical figures are 19–35 thousand flowers tonne^{-1}. Vase life varies slightly with cultivar, within the range 6–8 days under standard conditions at 15.5°C. Flowers are sensitive to ethylene, which shortens their vase life. The flowers exude mucilage into the vase water, and this can block the xylem vessels and impede water uptake by other flowers such as roses, carnations and tulips. It is best not to arrange other flowers with narcissi, but if unavoidable, keep the narcissi in a separate container for 24 h, then wash the stems before arranging the flowers. Recutting the stems produces more mucilage.

Pot Plant Production

Production of pot narcissi is similar to that for cut flowers. There are some differences in detail, as the cultivars used for cut-flower production are in general too tall to produce aesthetically acceptable pots, and require chemical dwarfing or some support to prevent the flowers and foliage falling over. A maximum desirable height is 25–35 cm at the goose-neck stage, and the most suitable containers are 14 cm half-pots or bulb bowls. Each should contain about ten flowers, produced from five or six large bulbs, planted in peat, tree bark or perlite–peat – many substrates have proved successful, provided the pH is above 4.0. Taller plants are produced by extra long low-temperature treatment, by low light in the glasshouse and by a low glasshouse forcing temperature. Ethephon can be used to dwarf standard commercial cultivars when they are 10–15 cm tall, or miniature cultivars can be used. These are mainly of the Cyclamineus types; cvs February Gold, Jack Snipe, Jumblie, Minnow, Peeping Tom and Tête-a-Tête are popular pot plants.

Pots are transferred directly into the glasshouse from the rooting room or standing ground on pallets, arranged at the highest pot density achievable. They should be marketed at the pencil stage of flower opening shortly after applying the dwarfing chemical; they should then open in 3–4 days. Costs of transport to market are higher than for cut flowers but the product has a higher value. For continuity of sales, consistent high quality and uniformity must be maintained.

Forcing Tazetta Narcissi

Unlike the majority of the narcissi that are grown commercially, those belonging to Division 8 of the RHS classification have no cold requirement for successful flowering, so their forcing is considerably simpler to achieve. The best known of this group are the multiflowered, scented, paperwhite narcissus (produced commercially in Israel and Malta), the cv. Grand Soleil d'Or (grown as an early outdoor crop on the Isles of Scilly) and the 'Sacred Lily' of China. Named cultivars bred in Israel include Sheleg and Ziva. Recommendations are for wholesalers to store bulbs on receipt in well ventilated stores at 25–30°C until shoot or root emergence is observed, then to transfer them to 2°C (as a 'holding' temperature to stop further development prior to housing, not as part of a cold requirement) followed by 2–3 weeks at 17°C before despatch to the forcer, who should plant the bulbs immediately on receipt. Treatment is similar to that described above for other narcissus for planting media, etc., and the growing temperature should be 16–17°C, as for the glasshouse phase of other narcissi.

It can take between 2 and 5 weeks to reach marketable stage (first floret fully coloured), development being more rapid as the forcing season progresses. The marketing season for cut flowers is late November–mid-April for 'Sheleg' and to the end of February for 'Ziva'. Potted plants are marketed at a slightly earlier stage than cut flowers – before the flowers open – and the plants are often treated with ethephon when shoots are 10 cm tall to restrict the final height. Recent research has shown that bulbs of cv. Ziva can be stored for long periods allowing year-round production of flowers.

FORCING OTHER 'BULBS' WITH A COLD REQUIREMENT

Many plants have a requirement for a warm–cool–warm sequence to achieve successful flowering, as already described above for narcissi. The temperature requirements of tulip are remarkably similar to those of the majority of narcissi.

Tulip Forcing

The commonly grown, and forced, tulip cultivars have been listed in Chapter 3. About 800 are currently grown, and the popularities of different cultivars wax and wane, as new ones appear and older ones go out of fashion or are superseded.

Some details have already been given of bulb treatment from lifting to the end of storage, but the broad outlines are that lifting date is determined by the subsequent treatment intended. For early forcing, early lifting is necessary, before the foliage has senesced, and the bulb tunic is beginning to turn yellow, usually in late June–early July. For mid-season forcing, lifting must be done by mid-July to allow cleaning and grading to end before the completion of flower differentiation (about mid-August), whilst the lifting date of bulbs intended for late forcing is least critical and follows the others in the latter half of July.

After lifting, bulbs are cleaned and graded, those for early forcing being dealt with immediately. Three bulb grades are used for forcing, the largest bulbs being used for the earliest rounds, and the smallest for late forcing. They are 12–13, 11–12 and 10–11 cm. Weights vary with cultivar differences in bulb shape, those for 'Apeldoorn' being, as kg 1000 bulbs^{-1} (also g bulb^{-1}), 27.8, 21.0 and 19.6, respectively.

Storage of large bulbs for earliest forcing at 35°C for 5 days (or 34°C for 7 days, more convenient practically), followed by 20°C, results in up to 17 days' earliness in reaching Stage G (and therefore the earliest date for starting low-temperature treatment) compared with bulbs stored at 20°C throughout this period.

A pre-planting storage period of 6 weeks at 9°C is usual for early and mid-season forced tulips, but for the late-forced ones 20°C is maintained until the end of August, then 17°C until planting, with low humidity in the store to prevent rooting. Soil and planting media should not be too coarse or too fine, should have an open structure with good porosity and water holding capacity, and a pH of 6–7. After planting, the bulbs should be covered with coarse sand to a depth of 1–2 cm to provide support for the growing plants. Although the bulbs are generally smaller, they are not planted as close together as narcissus. A mid-range average is about 17 000 bulbs per 100 m² of glasshouse area, including paths.

As with narcissi, post-boxing low-temperature treatment can be given in a controlled temperature store (double cooling), on a standing ground, or in a stack. Despite the extra costs of providing and running controlled temperature stores, the pre-housing phase of forcing is increasingly being done in such facilities rather than relying on ambient temperature outdoors. Such facilities also allow the use of more complex but efficient schedules such as introducing a short period at a lower temperature (e.g. 5°C) to satisfy fully the cold requirement in the latter part of the cooling period, after the bulbs are well rooted (which requires c. 9°C). The plants can also be 'held' at 0–2°C until housed. In this way crops can be accurately targeted to selling dates, or delays in clearing preceding crops from the glasshouse can be accommodated. Improved precision of temperature treatment and of crop timing helps prevent serious crop losses.

The total cold requirement (including that given pre-planting) to reach anthesis in 21 days in the glasshouse has been discussed under Physiology (Chapter 5); it varies with cultivar from 14 to 24 weeks. With accurate control of storage temperature and a knowledge of the cultivar's requirements, an estimate of housing date can be made by calendar, but for those treated on an outdoor site, a morphological assessment of the position of the flower bud relative to the tip of the bulb is necessary, as with narcissus. For rapid growth to anthesis, the base of the bud should be clear of the bulb tip, but for earlier anthesis, housing at an earlier stage is necessary, although this requires a longer period in the glasshouse.

Boxes are transferred to the glasshouse on pallets using a fork-lift truck, or via a roller conveyor, and placed on benches or directly on the ground, but isolated from direct contact for hygiene reasons by plastic sheeting (Fig. 8.4). The usual glasshouse temperature for tulip forcing is 18–20°C, and a relative humidity no higher than 80%. The temperature can be varied to meet marketing dates and lowered to lengthen stems and improve quality. In mild,

Fig. 8.4. A pallet load of tulip boxes (32 in all, eight high) just brought in from the cold store at the end of low-temperature treatment, for placing under the benches in the glasshouse. The shoots are yellow because the cold store is dark, but become green after a few days in the daylight. The thermal screen is in the extended position in this house.

muggy weather, it is often necessary to raise the temperature slightly or provide a little ventilation to avoid problems with *Botrytis* or topple (see Chapter 9). To speed throughput, a second batch of tulip boxes is often housed under the benches (Fig. 8.5) as with narcissus, and for the earliest forcings, covering the growing shoots with shade material such as old newspapers helps to increase stem length.

In theory, each tulip bulb should produce a flower, but some losses invariably occur. In the absence of disease or of problems from uneven watering, of inaccurate or uneven temperature treatment, or of rodent attack, all of which can affect part of a crop whilst it is in store or on a standing ground, a yield of over 95% can be expected. Flowers are harvested when the bud is half coloured, either by cutting with a knife at the bulb tip or by pulling up the plant with attached bulb. In the latter method the flower keeps better until bunched, usually into fives, and the bulb can be cut basally so that the

Fig. 8.5. Two 'rounds' of tulip bulbs in the glasshouse. That on the bench needs a further week before picking starts. When it has been picked, the boxes on the ground will be raised onto the benches, and a new batch from the cold store will replace it. The boxes are on the bench for 10–14 days at the start of the forcing season, and less than a week at the end. Each box is 60 × 45 cm and holds 125 bulbs.

flower stem is *c*. 2.5 cm longer, a useful addition for short stemmed cultivars or for early, or rapidly forced, crops which tend to produce short stems. The attached bulb also allows dry cut flowers to be stored before sending to market, at 0°C and high humidity, for 5–14 days longer than flowers cut conventionally. Before sale, tulips require 'hardening', when they take up water and there is some stem extension, particularly of the top internode of some cultivars. It is important therefore, to ensure that they are stored upright or they will set into curved postures. Normally 30–40 bunches make up a box.

In the USA, complex temperature sequences for forcing are often employed, using controlled temperature (CT) rooting rooms, which are necessary if winter temperatures are so cold that treating the bulbs outside would damage them, or, in the southern states, are not sufficiently cold to satisfy the cold requirement. Following pre-cooling (the 6 week pre-planting storage period at 9°C) if this is required for early flowering, the planted bulbs are kept at 9°C until roots emerge from the base of the container, and then go into 5°C. When shoots are 2.5 cm tall, a lower temperature (0–2°C) is used to retard further growth until the plants can be moved to the greenhouse, usually to meet a specific marketing date. For a range of cultivars and a succession of

Fig. 8.6. Beds of five-degree tulips being forced in an aluminium glasshouse. These have had their low-temperature treatment as 'dry' (i.e. unplanted) bulbs, which are then planted shallowly in 1.5 m wide beds in the sterilized border soil of the glasshouse at a high density, with 45 cm paths to allow picking.

flowering dates, at least two large controlled temperature stores and accurate scheduling of space usage is necessary. In recent years, similar recommendations have been made in The Netherlands for both standing grounds and rooting rooms, using lower temperatures as winter progresses; starting in October at 9°C, down to 7°C from 25 October, and to 5°C from 5 November.

A completely different system of tulip forcing called 'direct forcing' or 'five-degree forcing' provides the whole of the low-temperature treatment to dry (i.e. unplanted) bulbs. To avoid flowering failures, large bulbs of 12–13 cm grade are essential. Whilst the technique has been adopted by many of the best growers, it requires experience with the crop, the selection of suitable cultivars and considerable attention to detail to achieve success. At Stage G, or slightly later, the bulbs are stored at 5°C for 9–12 weeks then planted immediately into sterilized glasshouse soil at a density of 250–300 m^{-2} (Fig. 8.6). To avoid losses associated with poor rooting, it is necessary to remove the tough bulb tunics either before starting the cold treatment, or at planting. Bulbs are planted into soil which must be moist but not over-wet, and at a soil temperature of 9–10°C for the first 2 weeks and generally lower soil and glasshouse temperature than with conventional forcing (15–16°C being suitable). The time in the glasshouse is longer than with conventional forcing; 5 weeks is a common duration.

Autumn flowering of tulips, called ice tulips or Eskimo tulips satisfies a limited market. Only some cultivars are suitable. The bulbs are planted and allowed to root at 9°C, and are then stored at − 2 to − 3°C for 7–10 months, in a controlled temperature freezer. On removal, they must be thawed slowly; they take 14–18 days to force. Such tulips are expensive because of the cost of the treatment. This system is being expanded to provide near year-round flowering. Storage temperatures are lowered stepwise from 23°C in mid-September to 17°C before starting cooling at 9°C from mid-October to mid-December. The bulbs are then planted or packed in peat in polythene covered boxes and frozen at − 2°C until removed for forcing in glasshouses at 11–18°C.

Tulips make attractive pot plants, but for aesthetic reasons it is necessary either to use dwarf (genetically short) cultivars, such as the greigii hybrids, or to treat cultivars of normal height with a suitable dwarfing chemical such as ancymidol or paclobutrazol. Both methods have some drawbacks; dwarf cultivars are expensive, and the chemicals are not approved in all countries, are difficult to apply evenly and add to the final cost of the product. Some cultivars respond to a longer storage period before starting the cold treatment, and reducing the duration of low-temperature treatment by 1 week. Bulbs are planted three to a 10 cm pot, six to seven to a 15 cm pot or 10–12 to a 20 cm bulb bowl, with a well draining sterile medium of pH 6–7. If dwarfing compounds are to be used, it is best to avoid organic media, to which the chemical becomes adsorbed, reducing its efficacy. Planting the bulbs with their flat sides facing outwards ensures that the lowest, large leaves develop to the outside of the pot. Forcing procedures and temperatures are the same as used for cut flowers, and they are removed from the glasshouse for marketing at the green bud stage.

Vase life of cut tulips depends on cultivar, varying from 5–6 days for Apeldoorn to 11–12 days for Lustige Witwe in standard conditions of temperature and humidity to simulate domestic conditions. Despite some seasonal variations, there is no evidence that storage or forcing conditions have any measurable effect on vase life.

Iris Forcing

Dutch iris bulbs require no post-planting low-temperature treatment in rooting rooms for forcing. Immediately after harvest, bulbs intended for forcing are given a period of high temperature or treatment with ethylene to increase the rate of flower development and ensure a high percentage of flowering bulbs, followed by a treatment of 6 weeks at 9°C to promote early flowering. The exact high-temperature treatment applied depends to some extent on season and where the bulbs are grown; the lower the ambient temperature near lifting time, the greater is the amount of 'heat curing' required. Unfortunately there are no morphological indications of horticultural 'maturity' of the bulbs to aid the selection of appropriate treatments. Further, treatments given to bulbs are affected by transport, as bulbs grown in Washington State, Israel, France and The Netherlands are often forced hundreds or thousands of kilometres away.

In the USA, bulbs of Ideal intended for early forcing are harvested from mid-July to the end of August, stored immediately at 32°C (cured) for 10–15 days, graded, then kept at 18–20°C (stabilized) for 3 weeks. The stabilization often takes place whilst the bulbs are in transit from the grower to the wholesaler. Before planting the bulbs receive 6 weeks of 9 or 10°C (end treatment). Although this sequence has been used successfully for many years, experimental work indicates improvements from a sequence of 32°C for 4 weeks, then 2 weeks at 20°C and 6 weeks at 10°C for 10–11 cm bulbs harvested after 1 August and intended for early forcing (Doss, 1981). In contrast, the Dutch recommendations are; 35°C (2 weeks), 40°C (3 days), 20°C (3 weeks) and 9°C (6 weeks). Whilst these sequences are similar, the differences can be crucial in terms of percentage marketable flowers and the time taken to reach anthesis. Bulbs from Israel require much less high-temperature treatment, as do those intended for later forcing.

As with other forced bulbs, successful early forcing depends on the use of large bulb grades (10 cm and above) whilst 8–9 cm ones are acceptable for later forcing. Bulbs are planted in beds in the glasshouse soil, at densities of 140–250 m^{-2}, depending on bulb grade. Irises are more dependent on high light than narcissi and tulips, and there is a temperature–light interaction affecting flowering. Flowers abort ('blast') in low light at high temperatures, especially in the last 3 weeks before flowering, so in mid-winter it is advisable to use good quality glasshouses with high light transmission, low temperatures, and wider spacings, and to lower temperatures by 2–3°C during long spells of dull weather. At a glasshouse temperature of 15°C, the average time to anthesis of Ideal is 10 weeks. For the earliest flowering, by Christmas, it is therefore essential to use a higher temperature, 18°C, for the first 4 weeks after planting, followed by 15°C to avoid flower blasting during the susceptible second half of the glasshouse period.

The above are generalizations. Treatments for individual cultivars differ (Prof. Blaauw, for instance, is a later flowering cultivar than Ideal, and is not forced to flower before January), bulb sizes differ between cultivars, there is a beneficial effect of pre-lifting ethephon treatment or treatment with smoke or ethylene during storage producing a higher flowering percentage, earlier flowering, less blasting and better flower quality. With pre-lifting seasonal effects, differing temperature treatments and different forcing dates, it is not surprising that progress in improving forcing techniques is achieved empirically and iteratively, and that experience is a vital ingredient of successful iris forcing.

Retarded irises are an attractive spot crop because the bulbs are available all year and take only a few weeks to flower. Storage at 30°C results in such low respiration and metabolic rates that bulbs can be stored for a year or more – until the next season's bulbs are available. Treatment of the bulbs removed from store depends on the season and the subsequent growing conditions (forced, grown in the open). For glasshouse growing in the autumn, a common treatment is 9°C (6 weeks), 17°C (2 weeks) for Ideal and 9°C (8 weeks), 17°C (2 weeks) for Prof Blaauw. For forcing after mid-February, the recommendation for both cultivars is 17°C (2 weeks), 9°C (6 weeks), the reversed treatment stimulating leaf growth, delaying anthesis and producing tall, good quality flowers.

The best stage for picking is when the flower has fully emerged from the sheath and is showing 3 cm of colour for Ideal and 5 cm for Prof Blaauw, with the falls just beginning to separate. Cut flowers can be stored, but only for a few days, preferably in water, and at 0–1°C. Flower opening can be improved using flower preservatives containing silver thiosulphate.

The miniature irises *Iris reticulata* and *I. danfordiae* are suitable for forcing in pots. Unlike the Dutch irises, they have a cold requirement, similar to that of tulips, and the procedures followed are the same as those quoted above for the USA, using a controlled temperature rooting room. As the bulbs are small, it is usual to use 6 cm ones, planted as thickly as possible. Their requirement is for 15–16 weeks of cold. A glasshouse temperature of about 15°C is usual, and the pots are marketed when the sprouts are about 6 cm tall.

Hyacinth Forcing

After lifting from the field in mid-June, hyacinth bulbs intended for sale as 'prepared bulbs' are dried, cleaned and graded then stored at high temperature to promote early flower initiation. The largest inflorescences are produced by the largest bulbs, and 17–18 cm ones are generally used for early forcing. A typical schedule for earliest flowering is 30°C for 2 weeks, 25°C for 3 weeks followed by 23°C until the top florets are at the A2 stage of development, and then 17°C until planted. Those intended for later forcing receive less warm treatment. Whilst cultivars differ in the details of their optimum temperature requirements, these can be described broadly as a cold requirement of 8–11 weeks at 9°C followed by forcing in glasshouses at 20–25°C for the earliest flowers and 18 or even 16°C later in the season. To produce large, fasciated inflorescences, a treatment of 10 days at 23°C is given before the high temperature; this cannot be used for the earliest flowers because the 10 day delay in flowering is unacceptable. The consumer of purchased hyacinth bulbs aims to provide a sufficient duration of low temperature before bringing in the plants to room temperature to flower. Prepared bulbs in pots require 8–11 weeks (depending on cultivar) at 9°C before bringing them into forcing temperatures of 23–25°C and an r.h. of *c.* 86%. Covering the shoots at housing with black plastic prevents the shoots drying out and damaging the top florets. Cut flowers are forced at a lower temperature (18–22°C) to produce longer stems. For further details see Rees (1987).

Lily Forcing

Most of the lily bulbs forced in the UK are grown in The Netherlands, the silt soils of the UK bulb growing area around Spalding not being very suitable for lily bulb production.

The bulbs require cold treatment at 2°C, the normal durations being 6 weeks for Asiatic hybrids and 8 for the Oriental ones. It is also possible to use 'frozen-in' bulbs, which are kept at −1°C after their pre-cooling treatment, for year-round flowering. A well drained, sterile medium is required with a pH near 7. Planting density depends on cultivar, bulb size and time of year, within the range 27–90 bulbs m^{-2}, with more space being required for winter forcing, for large bulbs and for Oriental cultivars. Light is important. In summer, high light reduces stem length, and low light in winter leads to flower abortion and abscission. These disorders can be avoided by supplementary lighting with high-pressure sodium lamps SON/T at an input wattage of c. 500 per 10 m^2 for October, November and December plantings switched on all day until mid-March, except on bright days when they can be turned off between 10.00 and 14.00 h (Fig. 8.7).

A night temperature of 16°C is recommended for forcing, with the day temperature below 21°C, and the plants require regular liquid feeding or the use of controlled release fertilizers. Many cultivars require support because of the eventual height of the plants. Nets with a large mesh are recommended to avoid flower damage when picking. From planting to harvesting takes 8–10 weeks for Asiatic hybrids and 14–16 weeks for the Oriental hybrids, but frozen-in bulbs take less time. The period from the buds visible stage to picking is long; 30–35 days for Asiatic lilies and 50–55 for the Oriental ones. The blooms are cut when the first flower is fully coloured but not open (Fig. 8.8). Storage of cut lily flowers is best avoided, but they can be stored, dry at 0–1°C for up to 6 weeks in tightly sealed plastic bags following conditioning with silver thiosulphate, or wet, following conditioning, in containers with water at 0–1°C for up to 4 weeks.

Production of pot lilies is similar. The aim is a marketable plant 30–40 cm above the pot rim, using growth retardants if necessary. Large bulbs (16–18 cm) are planted singly, with the bulb set upright at the bottom of the pot. Small ones (10–12 cm) are set three to a 15 cm pot, and angling the bulbs towards the side improves the final appearance. Recommended forcing temperatures are 21°C by day and 13–17°C at night.

The production of pot Easter lilies, mainly cvs Nellie White (Fig. 8.9) and Ace is a challenge to growers in North America, to meet a demand for a well

Fig. 8.7. Lilies cv. Enchantment being forced in February in beds in border soil at 100 m^{-2}. The bulbs are planted by hand, 13 cm deep from both sides of the bed. Supplementary lighting is provided in winter by the 500 W high-pressure sodium lamps seen suspended above the crop.

Fig. 8.8. Bunching lily flowers. At harvest (when two to three buds are showing good colour in winter, and just colouring in summer) the whole stem is pulled up, and the bulb is cut off with secateurs and discarded. Lower leaves are stripped mechanically, the blooms are bunched in fives and packed 25 per box.

Fig. 8.9. A well-grown chemically dwarfed pot plant of Easter lily cv. Nellie White at point of sale. The 20 cm bulb was planted in a 15 cm pot, giving a total height of 55 cm.

grown product, of controlled height and maximum flower number on an annually variable date (Fig. 8.10). The bulbs are mostly produced in the coastal areas of Oregon and northern California, and when received by the forcer are at different stages of maturity depending on grower location and season. They are grown for sale in the few days before Easter, a feast which varies annually, the date of Palm Sunday falling, in the present century, between the extremes of 16 March and 18 April. The basic requirement for cold treatment for 6 weeks at 1–4°C can be satisfied by pre-cooling by the shipper or forcer, by natural cooling by storing the bulbs outside for 1000 h at temperatures preferably between 5 and 7°C, or by the use of controlled temperature rooms after planting, which is a more accurate and reliable method. After storage at 16–17°C for up to 4 weeks, when the shoots are

Fig. 8.10. Easter lilies in pots being forced in Michigan, USA. Note the overhead irrigation system and the lamps over the benches.

2.5 cm tall, the plants are exposed to 0.5–7.0°C for 6 weeks. The pots are then grown in a greenhouse at a temperature, at bulb depth, of 16–17°C. Three growing stages are recognized. The first is from planting to flower initiation (about 19 January), the second is to buds visible (on the first Sunday in Lent), and the third is a fixed period of 35 days from buds visible to marketing on Palm Sunday.

In Stage 1, long-day treatment can be applied as soon as sprouts emerge from the soil, to make up for any lack of maturity, to ensure that the cold treatment effect is fully satisfied, or as an insurance if Easter is very early. Stage 2 is of variable duration and may be as short as 4 weeks. Leaf number is fixed at the end of Stage 1 because flower initiation ends leaf production. In Stage 2, growth is controlled by leaf counting, on three representative plants from each batch of bulbs. On 28–31 January, the leaves yet to be unfolded from the apical tuft or spindle are counted, and their number divided by the number of days remaining before the first Sunday of Lent, to give the required 'unfolding rate'. A week later the process is repeated to determine the actual unfolding rate, and the glasshouse temperature is adjusted accordingly to ensure that all leaves will be unfolded by the due date (1.6 leaves unfolding per day is about correct). For Stage 3 a greenhouse temperature of 17°C will bring the plants to the desired state, with the flowers at the puffy white stage, for marketing on time. The temperature–development rate relationship has been modelled mathematically. For further information on Easter lilies, consult Roberts *et al.* (1985) and for other lilies MAFF (1984a) and Baardse (1977).

Potted Mixed Bulbs

A popular item for apartment-dwellers, and those without a garden or living in regions with hard winters, are attractive containers planted with several bulbs of different species, called 'Indoor Spring Gardens'. The growers provide the necessary cold treatment, the pots are sold from January to March, and the purchaser 'forces' the flowers in the living room of his/her home. Typically, a 20 cm bowl would contain six crocuses, six dwarf irises, a narcissus, three tulips and a hyacinth, with the taller plants at the back of the display. The purchaser would enjoy 2 weeks of growing, followed by 3 weeks of a sequence of flowering.

The temperature treatments are provided in a growing room, using a sequence of 9°C (4 weeks), then 5°C (4 or 8 weeks) and finally 0–2°C to a total cold period of 15 weeks, similar to the American treatment of tulips described above. The pots are marketed directly from the growing room.

Minor Bulbs

Many 'bulbs' plants are suitable for growing in pots. The following lists suitable subjects other than those described in detail in this chapter; those asterisked need cold: *Allium karataviense*, Anemone blanda*, Chionodoxa*, Crocus*, Eranthis cilicica*, Fritillaria meleagris*, Muscari armeniacum*, Ornithogalum thyrsoides, Oxalis adenophylla*, O. regnellii* and *Puschkinia**.

Only blue, white and bicoloured crocus are suitable for forcing in pots for Christmas, the yellow ones can be flowered from mid-January onwards. After a minimum of 14 weeks at 9°C they can be housed. Glasshouse temperature must be below 18°C, and they flower in 18–22 days.

Muscari require 15–16 weeks of cold at 5 or 9°C. To avoid long leaves which conceal the flowers, it is best to give most of the cold treatment before planting, which should be done not longer than 4–6 weeks before housing. There is no cold requirement for leaf growth, so they start to elongate at planting. For best colour development of the flowers, high light is necessary.

FLOWERING OF 'BULBS' WITH NO COLD REQUIREMENT

Freesia Flowering

Freesias are attractive, highly scented cut flowers with a colour range of yellow, cream, white, pink, red and blue, with both single and double forms, available year-round. They are a popular greenhouse crop in western Europe produced from either seed or corms, most of which are produced in The Netherlands.

The hard coated seeds need scarification and a 24 h soak in water at room temperature to ensure rapid and even germination. They are then planted in sterilized greenhouse beds using a sowing frame to ensure proper spacing within the range $125-150\,m^{-2}$, the lower densities being used for crops intended for winter and early spring production. Alternatively, they may be grown in pots, at about the same density, for cut-flower production. The modern tetraploid freesias which are most popularly grown have large seed, $1600-1800\,g^{-1}$, depending on cultivar. For good germination, a fairly high temperature of $18-22°C$ is required. Growing temperatures chosen depend on time of year and the target flowering period. For slow growth in winter, $10°C$ is adequate, but it is more rapid at $12-15°C$. Liquid feeds are normally applied at the buds visible stage and again 4–6 weeks later. Flower bud initiation occurs at the fifth leaf stage, and it is important to have developed a sturdy vegetative frame by this time to ensure strong stemmed flowers with many florets. Temperature at the time of flower initiation is critical. To avoid delayed initiation and floral abnormalities, soil temperatures should be $13-17°C$ during this period. Lower temperatures can result in too-early flowering and poor quality blooms, whilst initiation can be delayed for several months by high temperatures in summer.

Corms, after receiving their heat treatment to overcome dormancy, can be stored but not for longer than 4 weeks from the end of the heat treatment. They are planted at $65-110\,m^{-2}$, depending on season and cultivar, quite shallowly (unless warm surface layers make a deeper planting necessary) because the contractile roots will pull down the corms. Recommended growing temperatures are $12-15°C$ in winter and $15-20°C$ in summer, but these are not critical until near flower initiation, which starts about 6 weeks after the end of heat treatment.

Flowers are harvested with a knife, when the first floret is just opening and two others are showing colour, the position of the cut depending on the market. In the absence of a premium for long stems, the cut is made above the first side shoot down the stem (about 20 cm), but if necessary, as in Holland, a longer stem (40 cm) may be achieved with a lower cut, and the side shoot(s) trimmed at bunching. Flowers are bunched in fives, usually of mixed colours in a tapered transparent sleeve, and the bunches are boxed, end-to-end in bundles of five giving a total of 200–250 per box. The cut ends are often wrapped in a moistened pad for transport to market. If necessary, they can be stored dry at $0-0.5°C$ for 7 days or at $9-10°C$ for 5 days. The flowers are sensitive to ethylene and their life can be extended by flower preservatives. Growing freesias for sale as pot plants has so far not been highly successful, with variable results. Pre-planting temperature treatment has been reasonably successful with some cultivars, and there has been some success with dwarfing chemicals. High light and low growing temperatures give the best results. It is expected that improved methods will soon be developed in The Netherlands and the USA. For more information on freesias see Smith (1979).

Alstroemeria Flowering

In recent years, *Alstroemeria* hybrids have become a popular and important greenhouse cut flower among the top ten of all cut flowers in The Netherlands auction turnover values. New hybrids are being produced, extending the flowering season and widening the colour range of bronze, orange, pink, red, yellow and white. Planting material, obtained from approved specialist propagators under licence, comprises rhizomes propagated by division, or, more usually, grown-on after tissue culture. Rhizomes are normally planted between June and August, directly into beds 1 m wide in the greenhouse. A double row system per bed is usual, with spacings of 40 × 50 or 50 × 60 cm, depending on cultivar. Temperature is maintained at 15°C until the plants are established, then at 10–13°C. An advantage of the crop is that it does not require expensive high temperatures, and it can tolerate winter temperatures as low as 5°C. In summer it is advisable to keep soil temperatures below 20°C, and there is experimental evidence that a day temperature of *c.* 20°C and a soil temperature of 12–14°C is near the optimum for continuous production. Because plants grow tall, support is necessary, and three to four layers of nets are provided at planting time. Periodically, weak and non-flowering shoots are removed, especially in late autumn and winter. Flowering can start in the autumn or be delayed until the following spring, depending on cultivar. Flower production is low in winter, increasing as light increases, but can be improved by supplementary lighting or a treatment providing a 13 h photoperiod, which results in early flower initiation.

Flowers are cut when the first flower is fully opened, and four to five florets are open in the inflorescence. The flowers are sensitive to ethylene and the leaves are fragile and turn yellow after harvest. Storage at 4°C in water is successful for 2–3 days, but dry storage is not recommended.

Zantedeschia Flowering

Several species of this genus, also called arum lilies or calla lilies, are produced as cut flowers (*Z. aethiopica* the large white arum lily with flowers 45–60 cm tall) or as pot plants (*Z. albomaculata, Z. rehmanii* and several hybrids coloured pink, yellow or lavender, some with spotted leaves, and about 30–45 cm tall). Rhizomes of *Z. aethiopica* are produced in Israel and The Netherlands, and tubers of the other, smaller plants, from France, The Netherlands and New

Zealand. Several methods of growing are practised, using raised greenhouse beds or pots for cut flowers as well as pot-plant production, planting densities being from 10 × 10 to 50 × 60 cm, depending on tuber size, or three tubers to a 20 cm pot. For pot-plant sales, tubers are planted singly in 10 or 12 cm pots. Day/night growing temperatures recommended are about 20/13°C for the white and pale yellow types or 20/16°C for the golden yellow and pink ones. It is normal to provide shade in summer because flower quality is reduced by high light. The flowering season extends from early winter to late spring. Flowers are cut, rather than pulled, when the spathe is fully open, before the start of pollen shedding; they can be stored dry at 4–5°C for up to 3 days, or wet at the same temperature for a week.

Flowering of Other 'Bulbs'

Some 'bulb' plants are grown from seed in much the same way as are other seed-propagated plants, where success depends on good husbandry and appropriate horticultural practice. A good example is cyclamen, grown commercially for sale as flowering pot plants. These are almost invariably grown from seed, except for the occasional need to propagate specific clones vegetatively, by corm splitting, for maintaining breeding stocks. Leaves will not root, and the corms grow in size annually, and do not produce daughter corms which would facilitate natural clonal propagation. Growth from seed is slow, with a gradual increase in corm and cotyledon size, and the first true leaves not appearing until *c.* 80 days from sowing. When there are six to seven unfolded leaves, the plants are potted up. Growing temperatures of 17–18°C at night and 20–24°C by day are recommended, it being important not to exceed these temperatures, and to provide shade in summer to encourage flowering. Flower buds initiate in the axils of the sixth and successive leaves, but flower bud development is very slow, the first flower reaching anthesis when there are *c.* 35 leaves unfolded, each with its axillary flower. In good growing conditions, plants last a long time. Individual flowers last a month, and the plant should keep at least 6 weeks in the home (Widmer, 1980).

For some other 'bulb' plants, flowering is achieved by the customer, who buys 'bulbs' and plants them for flowering indoors, in pots, or outside in the garden, in window boxes or hanging baskets. Many of these uses are long established, like hyacinths and hippeastrum for indoor growing. Full instructions are generally provided for the purchaser to ensure that the product performs satisfactorily. In other cases, such as cyclamen, the potted plants are sold in full flower, having been brought to this state by the grower. Aspects of outdoor planting have already been considered in Chapter 2.

OUTDOOR FLOWERING

Demand for flowers is greatest at times other than the natural flowering season. Many potential flower purchasers have flowers available in their own gardens in spring and summer, but there are situations where it is still commercially viable to pick natural season outdoor flowers for sale. There are holidays and festivals with a large demand for flowers, such as Mothering Sunday and Easter. Growers with saleable flowers will pick from the field any flowers that would help meet this demand. Growers in areas of mild climate, such as the Southwest and the Isles of Scilly, regularly grow outdoor 'bulb' flowers for sale in colder parts. Narcissi are picked and bunched in the field at the fat goose-neck stage to preserve high quality and to avoid the flowers being damaged by wind and rain. Unlike forced flowers, no leaves are picked, because this would reduce subsequent bulb yield, by reducing the photosynthetic area. Similarly, the coincidence of a market demand and available flowers encourages the picking of tulip flowers in the field from crops being grown for bulb production. It is important that one or two leaves are allowed to remain on the plant to ensure continued 'bulb' growth. For tulip, studies have shown that, compared with deheading, picking flowers with one leaf reduced bulb yield by 5–7%, whilst picking flowers so as to leave one leaf reduced it by 21–23%, averaged over six cultivars and two seasons.

As well as natural season flowering, bulbs with a cold requirement can be pre-cooled at 9°C for up to 6 weeks before planting to increase earliness of flowering in the UK by 3–4 weeks, a general technique that enables narcissi and tulips to flower outdoors in climatic areas which would not naturally provide sufficient cold for normal development (like southern Greece and California). Tazetta narcissus on the Isles of Scilly can be made to flower earlier by early lifting and warm storing the bulbs at 27°C before planting, but this technique does not give consistent results in practice, with considerable variation between years apparently because of differences in soil moisture.

Covering the soil surface with polythene sheeting to improve flowering is a successful technique widely used on the Isles of Scilly for tazetta narcissi. The cv. Soleil d'Or normally flowers from mid-January until late February, but with present techniques, the first flowers appear on the market in mid-October. The clear polythene is applied for 4–6 weeks from mid-May, and the crop is then defoliated after the sheets are removed in July. It is even more effective if combined with 'burning-over', a technique whereby straw or propane gas is burned over the soil surface. The two procedures modify the soil temperature and probably provide ethylene which has a physiological effect on the bulbs (see Chapter 5). Flowering of tazetta narcissi can also be retarded by covering the crop with clear polythene in mid-July, when the soil is hot and dry, and after the removal of dead foliage and weeds. The polythene is removed in October when the leaves start to emerge. Whilst retarding flowering is less beneficial than producing early flowers, it has the advantage of spreading the picking load and providing flowers over a longer period.

Anemone Flowering

Anemones are grown in southwestern England as an outdoor winter crop, flowering in late spring and early summer, and elsewhere in the UK as a minor protected crop under glass or plastic with minimum heat or none. In the open, good soil preparation is necessary, with an optimum pH of 6.5. Whilst a good potassium status is necessary, excessive nitrogen must be avoided because it leads to 'soft' plants and frost damage. Corms of 2–3 cm are normally used, at a depth of 5 cm in rows 45–60 cm apart and a density of 150 000–200 000 ha^{-2}. Planted in May–July, these start to flower in autumn with a peak of production in November–January. From seed, flowering is later and the duration of the crop is extended, stems are shorter, but flower colour is better. Seeds are sown directly into shallow drills, 45–60 cm apart, at about 1.5 kg ha^{-1}. Under protection, the recommended temperatures are 16°C by day and 7–10°C at night, and avoiding long periods above 16°C. When the petals are fully coloured the flowers are cut and bunched in fives or tens. They can be stored, dry, at 4–7°C for 4–7 days. Further information on anemones is available in MAFF (1977).

Gladiolus Flowering

Gladioli are planted after the danger of severe frost has passed. In some areas with mild winters, frost-free or almost so, they can be grown as a winter crop (in Mexico, Florida, Israel, Queensland and southern Japan), but at higher latitudes they are progressively grown as later, or summer crops. All cultivars except the Nanus and Colvillei types were selected for summer growing and are not well adapted to flower in short days. Flowering is 1–3 weeks earlier than in long days, spikes are shorter and bear fewer florets. It pays therefore to use the largest corms for winter growing, and some growers use artificial light to extend the day length.

Summer production of the grandiflorus, primulinus and butterfly gladioli is not difficult, but windy situations are best avoided because they are tall plants which grow best in full sun, in deep, well drained soils of pH 6–6.5. Planting depth is 16 cm to the corm base, at densities of c. 300 000 ha^{-1} for 8 cm corms, usually in rows 13–17 m^{-1} of row. The flowers are harvested 60–100 days after planting the corms; they are cut when in tight bud with one to five florets showing colour. Two or three leaves are left on the plant for continued corm growth. The flowers must be kept upright to prevent bending, and can be stored for 5–8 days dry at 4–5°C. Conditioning with silver thiosulphate and treatment to prevent *Botrytis* attack allows them to be stored dry for 2–3 weeks.

9

PESTS, DISEASES AND DISORDERS

Most 'bulb' plants are monocotyledonous and lack some of the protective mechanisms found in dicotyledonous plants. Despite the existence in some species of stem thickening and secondary growth, cambia are generally absent in the monocotyledons, so that the isolation of damaged areas and wound healing capabilities resulting from cambial activity, and especially suberization, are rare. There is a general lack of cell division in wounded parts of most monocotyledons, and even when this does occur, meristematic activity is extremely limited. Reconstitution phenomena are restricted to dermal and cortical tissues such as renewed epidermis and cuticle, exodermis or the sclerification of existing cells.

The generally high moisture content of the monocotyledonous plant body also contributes to vulnerability. In compensation for this, perhaps, bulb plants exhibit a range of chemical protective methods, many being unpalatable or poisonous. Paradoxically, many of these chemicals add to the attractiveness of the plants for human food (such as garlic, onions and their allies). Narcissus (as well as other amaryllids) contain many chemicals toxic to mammals, including oxalic acid and at least 15 alkaloids. These compounds and allied precursors have antimitotic properties, and narcissus plants have long been known to be toxic because of their narcotic, purgative and emetic properties when eaten. Their sap is a powerful irritant which can cause rashes on the hands and faces of flower pickers. Tulip bulb tunics contain tulipalin (α-methylenebutyrolactone), a barrier to fungal attack which causes dermal irritation to susceptible bulb handlers, colchicums contain colchicine which interferes with chromosomal behaviour in mitosis, and fruits of lily-of-the-valley are toxic to humans. Bulbs of *Urginea*, the sea squill, have for centuries been used for treating heart diseases. Sheep are frequently poisoned by eating anemone leaves, and day lilies are used medicinally in China to treat human schistosomiasis, although overdosing can cause blindness and death. Such biologically active constituents presumably have a natural role in protecting the plants from pests and diseases.

168

The vegetative propagation of so many 'bulbs' allows the spread of virus diseases throughout most bulb stocks unless specific steps are taken to establish a nucleus of virus-free material as a base for releasing certified material to growers.

Some fungal diseases affect many species; others are more specific. *Fusarium* bulb rots are common worldwide on a range of host species, but although the genus, and many of the species, e.g. *Fusarium oxysporum*, are ubiquitous, there are *forma speciales* associated with individual host species. In the same way, whilst some virus diseases have a wide host range (e.g. cucumber mosaic virus), others are restricted to single species. The above gives an idea of the breadth of this subject. Detailed treatment of the individual pest and disease problems of all 'bulb' plants is outside the scope of this book; the reader is referred to Alford (1991), Brunt *et al.* (1979), Lane (1984), Gratwick and Southey (1986), Linfield and Lole (1989) and Scopes and Stables (1989).

Rather than deal with specific pests and diseases of individual species, it is convenient to survey in more general terms the different types of major pathological problems, by kind of pathogen (fungal, viral, insect, nematode, etc.), the organs attacked (foliar, storage organ, systemic), and to outline possible control measures. Current legislation about the use of pesticides, the decreasing use of chemical control methods, and increasing pressures for the use of environmentally friendly methods make it difficult to recommend pest and disease control methods. These can rapidly become out of date because of the non-availability of the chemicals, or the withdrawal of approval for the use of a chemical on a certain crop. Its use then becomes illegal. Up to date information on chemicals approved for 'bulb' crops is available in the current issue of *The UK Pesticide Guide*, published jointly by CAB International and the British Crop Protection Council.

PESTS

Pests range from barely visible mites and nematodes to mammals and birds, with the severity of the resulting problem often in inverse relation to the size of the pest.

Nematode Pests

These can be among the most serious pests of major crops, requiring expensive control measures. In addition to the damage they cause directly, many are of importance as vectors of virus diseases. There are a number of species, divided into groups, reflecting the host parts attacked.

1. Leaf nematodes attack buds, leaves and plant apices of a range of

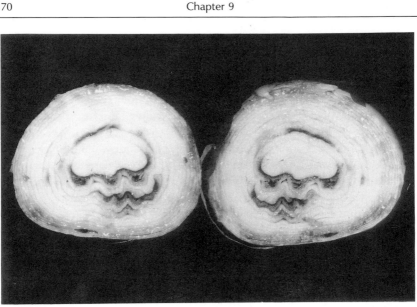

Fig. 9.1. Cross-section of narcissus bulb infested with stem nematode (*Ditylenchus dipsaci*) showing the characteristic concentric brown rings. (Photograph courtesy of *Horticulture Research International*, the copyright holder.)

glasshouse plants, including those being propagated. *Aphalenchoides* spp. have been recorded as damaging alliums, anemone, begonia, colchicum, crocus, iris, lily-of-the-valley, narcissus, scilla and tulip.

2. Cyst nematodes affect mainly the roots of herbaceous plants but are not serious for 'bulbs'.

3. Migratory nematodes (including root lesion nematodes) of the genera *Longidorus, Pratylenchus, Trichodorus* and *Xiphinema* can be important for a range of 'bulbs' in light, sandy soils and in protected cultivation.

4. Root knot nematodes especially those of the genus *Meloidogyne* affect pot plants grown in glasshouses such as begonia, cyclamen and gloxinia.

5. Stem and tuber nematodes are major pests of most commercially grown 'bulbs'.

The stem nematode of narcissus (*Ditylenchus dipsaci*) produces symptoms on foliage and bulbs. Leaves of affected plants are short, pale and often distorted with a few elongated lesions called spickels which sometimes contain eelworms. These often coalesce to produce necrotic patches, and the leaf rots. Bulbs are soft, with a dull appearance at lifting, and when cut across, show characteristic concentric brown rings (Fig. 9.1). Adult eelworms are 1.0–1.3 mm long, and only a few individuals are necessary to produce severe symptoms. During a growing season, they multiply to several thousand, congregate in the leaf bases, and migrate into the soil when the bulb dies.

Other plants can act as hosts until narcissi are replanted. It is essential that stocks are regularly inspected when growing, with suspect plants being lifted and destroyed, and the land spot-treated with nematicide. At least 5 years should elapse between bulb crops, with steps taken to destroy groundkeepers, other susceptible host plants (including onions and strawberries) or susceptible weeds (like bluebells).

At lifting, all bulbs should be inspected and soft or suspect bulbs destroyed. Stocks should be given hot-water treatment (h.w.t.) after lifting (every 2 years) at 44.4°C for 3 h with added formalin (0.2% commercial formalin plus wetter). This treatment has been in use for over 70 years, with some refinement with time to overcome some of the early difficulties because of primitive temperature control, uncertainties about timing of treatment, and the need to provide pre-treatment to avoid flower damage and to ensure effective eelworm kill. Nowadays, warm storage for 1 week at 30°C before h.w.t. is standard treatment ('pre-warming') to prevent subsequent flower damage, and timing of the h.w.t. is not critical within the period mid-June–August, irrespective of the stage of flower development within the bulb. As pre-warming tends to make eelworms more heat resistant, the bulbs are pre-soaked in cold water plus formalin and wetter overnight before h.w.t., and the temperature of the h.w.t. is raised to 46–46.7°C. However, stocks known to be infested should not be pre-warmed, as this lowers the efficacy of control; the flower crop the next year is sacrificed. These measures, if carried out carefully, will ensure control of the pest, despite the small margin between killing the pest and damaging the bulb.

The h.w.t. of narcissus bulbs requires considerable investment in equipment to handle mechanically large quantities of bulbs, allied to efficient systems of bulb handling, and layouts to ensure a one-way flow of bulbs through the treatment system to a 'clean' area to obviate possible re-infestation (see Gratwick and Southey, 1986). Tanks must be sufficiently large to contain at least twice the weight of water as of the bulbs to be treated in each batch, i.e. 2000, preferably 3000 l, per tonne of bulbs. Circulating pumps are required with a capacity of at least five changes per hour, with a working pressure of c. 1.5 kPa, and filters are necessary to remove bulb debris. The water can be heated directly by gas or oil burners, or using water from a steam calorifier, but electrical heaters with accurate and sensitive thermostats are usually used for fine temperature control. The tanks are of two types, either top-loading or front-loading (see Fig. 7.12). The former requires an overhead hoist, and the latter is filled by fork-lift truck, using palletized nets, palletized trays or bulk bins.

The tulip race of the stem nematode is particularly virulent; it attacks both narcissi and tulips, but the narcissus race cannot multiply in tulips. Symptoms in tulip are soft discoloured bulbs, sometimes with brown rings as in narcissus, and are more obvious above ground with scarred, split and bent stems, which die prematurely, and green streaked or poorly coloured flowers.

Hot water treatment of tulips is much less satisfactory than with narcissus because of plant damage. Treatment with thionazin was formerly widespread, but this chemical is no longer approved. Control is now achieved by careful stock inspection, removing and destroying diseased plants and applying rigorous hygiene and crop rotation. Similarly, there are no recommendations for effectively controlling the pest in snowdrops, where symptoms resemble those in narcissus; but with hyacinths, Dutch recommendations are for 4 h h.w.t. at 45°C.

Other nematodes are generally of lesser importance, although they can cause serious problems in individual crops. They include the potato tuber nematode (*Ditylenchus destructor*) which damages the storage organs of dahlia, gladiolus, iris and tulip. Symptoms occur only on the bulbs or corms, which bear dry, brown or black lesions, shoots are weakened and secondary infections occur. Control is by h.w.t. as for narcissus, using pre-warming and a pre-soak. In Cornwall and on the Isles of Scilly, the root lesion nematode (*Pratylenchus* spp.) feeds on roots, allowing fungal entry, and producing patches of poor crop growth in the field. The recommended treatment for these very small nematodes is field treatment with nematicides or general soil fumigants. An eelworm root-rot of hippeastrum can be controlled by a 2 h h.w.t. at temperatures between 40 and 50°C. At 47°C and above, the first flower is lost, but subsequent ones are normal.

Insect and Mite Pests

Bulb flies

There are three species of narcissus flies: the large narcissus fly, which can also damage bulbs of hyacinth, iris, snowdrop and others (*Merodon equestris*), and two small narcissus flies (*Eumerus strigatus* and *E. tuberculatus*). Of the small narcissus flies, the former attacks mainly narcissus bulbs, whilst the latter has a wider host range, including hyacinth, iris, lily, as well as many vegetable crops. All three bulb fly species are widely distributed in Europe and North America, and probably originated in southern Europe. The large narcissus fly is locally an important pest, is *c.* 12 mm long and looks like a small bumble bee. The adults fly in April–June, and lay single eggs on or near the surface of the bulb. These hatch, and the larvae enter through the base plate and feed on the scales until they reach a length of about 18 mm and pupate the following spring, finally emerging as adults. Large bulbs often survive, but may have leaves only on one side.

The small narcissus flies are only half the size of the large ones, and look like large houseflies when they first emerge in April. Eggs are laid in bulbs damaged by other means, often those infested with nematodes, and develop into larvae up to 9 mm long. As a secondary pest, control of the primary damage will reduce the numbers of infestations from the small narcissus flies. Several eggs are laid in each bulb and there are a number of generations in a season. Some control results from the removal of foliage and the mechanical closure of the hole left above each bulb by the dying foliage, whilst early lifting shortens the duration of exposure of the bulbs to the flies. Damaged bulbs should be removed and destroyed at bulb cleaning, and other inspections. Flies are killed by normal h.w.t. Bulbs can also be dipped in an approved insecticide prior to planting.

Aphids

Aphids attack many 'bulb' plants, either foliage and flowers in the field or glasshouse, or the storage organs in store. Aphids are also important vectors of virus diseases even if the plants are not colonized, as test probes of non-host tissue can transfer some virus particles and cause infection. Several species of aphid are involved. Growing crops of anemone, gladiolus, hyacinth, iris, lily, snowdrop and tulip can be infested, sometimes severely, by several species, including *Myzus persicae*, *Aulacorthum circumflexum*, *Macrosiphum euphorbiae*, *Myzus ascalonicus*, *Aphis fabae* and *Aphis newtoni*. Direct damage to foliage and flowers can reduce flower quality and lower bulb yields. Under glass, *Aphis frangulae gossypii* is common on ornamentals, including begonia and arum lily. In store, the commonest aphid is *Dysaphis tulipae* on tulip, crocus, gladiolus, iris, lily, and others. Emergent shoots can be distorted, and the quality of the flowers reduced to the point of their being unsaleable. In the field, spraying is advocated as soon as infestations are seen, whilst the use of insecticidal sprays or fumigants is recommended for stores.

Thrips

Gladioli can be badly affected by thrips (*Thrips simplex*). These small thin insects, yellow when young and dark brown when adult, cause streaky stems by their sap sucking. Affected flowers have small white streaks. The adults overwinter in store, where they continue to multiply at temperatures above 10°C, under the dry outer scales of the corms. Not easily noticed, they produce rough grey-brown patches on the corm surface. Sprays in the field and fumigation of stores are the recommended treatments. The same species also attacks crocus, freesia, iris and lily. A different species attacks lilies in store (*Liothrips vaneeckei*), but the same control measures apply. Western flower thrips (*Frankliniella occidentalis*) is a relatively recent accidental introduction into the UK which infests several glasshouse ornamentals including cyclamen and begonia, as does *Thrips tabaci*, whose host range also includes arum lily and dahlia.

Caterpillars

Caterpillars and cutworms are not generally important pests of 'bulb' crops, but those of several species, including the angle shades moth (*Phlogophora meticulosa*) eat the flowers and leaves of anemones and gladioli. The garden swift moth (*Hepialus lupulinus*) and its close relative the ghost swift moth (*H. humuli*) are common pests of ornamentals; their larvae can seriously damage the roots and storage organs of anemone, dahlia, gladiolus, iris, lily-of-the-valley and narcissus. Cutworms (caterpillars of noctuid moths like the turnip moth (*Agrotis segetum*)) can be troublesome on anemone, dahlia, gladiolus and iris in dry late summers, attacking plants at or just below the soil surface. Sprays and drenches, as appropriate, are recommended.

Tarsonemid mites

Several of these extremely small (*c.* 0.2 mm long) mites are important pests of 'bulb' plants. The cyclamen mite (*Phytonemus pallidus*) is a serious problem for growers of glasshouse ornamentals, including begonia, cyclamen and many aroids, whilst the bulb scale mite (*Steneotarsonemus laticeps*) affects narcissus and other amaryllids, including eucharis, hippeastrum, sprekelia and vallota. Affected broad leaved plants have contorted, crinkled foliage because damage to growing leaf margins prevents normal leaf expansion. This can lead to death of organs or the whole plant if infestation is severe. The bulb scale mites multiply rapidly throughout the year in the spaces between scales in the neck of the bulb, their feeding producing brown scars on the more angular part of the bulb scales. In forced bulbs, most injury is caused when the bulbs are housed, and rapid feeding by the mites coincides with scape and leaf extension. Damage to leaf edges and the scape angles, often producing a saw-edge appearance, results in these organs being deformed, and the flower bud can be killed. In hippeastrum, characteristic symptoms are red streaks and blotches on the growing leaves, often accompanied by mal-formed flowers. Treatment with specific acaricides is recommended, and repeat applications may be necessary for effective control. The bulb scale mite can be controlled almost completely by h.w.t. at 43°C for 1 h, and this might suffice for a crop to be forced. Complete eradication requires 3 h at 44°C, but there is an attendant risk of flower damage, so this is best reserved for a stock intended for replanting. Fumigation of dormant bulbs is effective, and a wise precaution for forced bulbs suspected of being infested is to drench them with dilute acaricide immediately after boxing.

Another mite which affects many 'bulbs' as a secondary pest is the bulb mite (*Rhizoglyphus* spp.). These are sufficiently large (0.7 mm) to be visible to the naked eye, and affect freesia, gladiolus, hyacinth, lily, narcissus and tulip. They increase the amount of original damage, even to the extent of killing the storage organ, the internal tissue becoming dry and powdery. Soil sterilization, 'bulb' dips and drenches and the use of systemic acaricidal granules are all advocated, as well as attention to the primary cause of the initial damage.

The broad mite (*Polyphagotarsonemus latus*) requires high temperatures, and is a pest of glasshouse ornamentals like begonia and gloxinia, feeding on the undersides of leaves and flower buds and distorting flowers and foliage, and in extreme cases killing the plants. Application of acaricide as soon as damage is seen, with repeated treatment when required, is the standard control measure.

Weevils

The vine weevil is a major pest of glasshouse pot plants and outdoor ornamentals, which has become more serious in recent years, possibly because of greater use of peat based growing media. Outdoors, eggs laid by the adult beetles in early summer hatch within a few weeks to produce large (8–10 mm) larvae which attack underground parts over several months until they pupate the following spring. In heated glasshouses, there is often a peak in adult numbers in the autumn, and the active period is extended. Affected plants wilt, often collapse suddenly and are frequently killed. Chemical control is best done by treatments incorporated into the growing medium, and there are effective biological control agents.

Other Pests

Many other pests of 'bulb' plants are well known horticulturally because of their wide host range. Others are of minor importance because they have a restricted distribution geographically, or a limited host range. In the first category are whiteflies like the glasshouse whitefly (*Trialeurodes vaporariorum*), controlled by insecticidal treatment and by biological control using the parasite *Encarsia formosa*, and the spider mites such as *Tetranychus urticae*, the two-spotted spider mite, a widespread pest of many outdoor and glasshouse crops. Treatment with a non-phytotoxic acaricide is recommended, but there are increasing problems of resistance. Alternatively, biological control can be tried, using the predatory mite *Phytoseiulus persimilis*.

Several species of slugs and snails are ubiquitous and damaging to many plants, especially at emergence or at the small seedling stage; keeled slugs attack bulbs of tulip and lily underground, where their activities can go unsuspected. Treatment with molluscicides is recommended, but the success rate is often disappointing. A coarse grit around susceptible plants can help, and bait, protected from rain, can supplement the use of molluscicidal drenches.

Occasional damage is attributable to beetles, such as the garden click beetle, *Athous haemorrhoidalis*, which is a common pest of all herbaceous plants, and growers find the lily bettle an important pest of lilies which also affects fritillaries. Although local and uncommon, it is becoming more serious as lily growing is increasing in the UK. The adult is 6–8 mm long with bright red thorax and elytra. Eggs laid in spring produce numbers of black larvae which feed voraciously, pupate below ground and reappear as adults, providing two generations a year.

Birds and small mammals, especially rodents, are locally destructive. Tulip bulbs and crocus corms are particularly attractive to the long-tailed field mouse (*Apodemus sylvaticus*), and squirrels dig up these too. Pheasants are particularly fond of tulip bulbs.

DISEASES

Diseases caused by fungi, bacteria and viruses can affect the whole growing plant in a systemic manner, as with viruses, or the whole plant in the 'dormant' condition, i.e. the storage organ(s), or affect only part of the growing plant, such as the leaves, or only the underground parts. Symptom expression does not necessarily reflect the site of the attack, as below-ground damage can result in the wilting or collapse of apparently healthy above-ground parts, or secondary infections of shoots and flowers may result from a bulb or corm infection. Control measures can involve hygiene to prevent infection, direct treatment aimed at killing the pathogen by chemical or physical means, or treatment which favours the growth of the plant rather than the pathogen, by such means as control of temperature or humidity. In many cases, a knowledge of the pathogen's life history can reveal times when pathogen growth or disease spread can be reduced by simple measures (e.g. spore dispersal dependent on water films can be prevented by dry conditions, and the growth of fungi requiring high temperatures can be restricted by late planting in the autumn).

Because of the clonal nature of so many 'bulb' crops, it is important to monitor the health of stocks, and to rogue and destroy any deviants, plants not true-to-type or those showing disease symptoms, irrespective of the severity of damage or the numbers of plants involved. The high cost of labour and increasing use of mechanized cultivation means that growing crops now have less personal attention from the grower and his staff, and the crop health is generally less good because of this. In compensation, more is known about the pathogens, and better control methods are available. Ecological considerations and pressures from the green lobby are restricting the use of many chemicals efficient in crop protection, and there is increasing pressure to develop effective physical and biological methods of control. For virus diseases, control of vectors and roguing are important, but the main thrust must be the production of virus-tested mother stocks, kept and propagated under conditions where they are virus free. These stocks are a source of 'clean' material for release to growers. As these become infected in field conditions, they are replaced periodically from the mother stocks.

International and local trade in 'bulbs' has been in progress for centuries, so it is not surprising that most diseases are common to all the 'bulb' growing areas, following distribution before effective plant health regulations were devised. There are some climatic effects, with *Fusarium* diseases more common where temperatures are high, whether this is because of latitude, or due to a more continental climate or to seasons warmer than normal. Leaf diseases are generally more serious in moist, more maritime climates than in drier areas.

Bacterial Diseases

A few bacterial diseases of 'bulbs' are serious, that of hyacinth, called 'yellow disease' (*geelziek* in Dutch) of hyacinth, caused by *Xanthomonas hyacinthi* being the best known. The bulb rots completely before planting or soon afterwards, so that there is no emergent shoot or only a diseased, weak, non-flowering one. Early in infection a transverse section of a bulb reveals small yellow spots, which ooze a sticky, bacteria-laden slime. The spots are elongated longitudinally, extend downwards to the base plate and then infect healthy scales and the shoot. Spread is rapid within the bulb, and secondary infections help the rotting. Spread to other plants is by wind and rain splash to neighbouring plants, and entry is effected via wounds. Suspected stocks are stored at high temperatures (30°C initially, 37°C later) to encourage the rot and facilitate the identification of infected bulbs, which are then destroyed. In the field, close scrutiny identifies infected plants at an early stage and helps prevent spread.

At flowering, *Erwinia carotovora* attack of hyacinths rots the scape at ground level, the shoot topples over and there is a rapid and malodorous collapse of the whole plant. The disease is called soft rot, and the same name is used for the very common bacterial decay of rhizomatous iris, caused by the same pathogen, which has also been recorded on freesia corms. Removal of the affected plants and sterilization of the soil are the only control measures practicable.

Tulips are sometimes infected with yellow pock (*Corynebacterium* sp.), which produces yellow spots on the bulb scales and a yellow staining of the vascular bundles as seen in transverse section. Severe infections kill the bulbs, and no shoot is produced. In less severe cases, stunted shoots bear characteristic silvery streaks; these plants die prematurely.

Gladioli worldwide suffer from bacterial scab (*Pseudomonas marginata*), which causes lesions on the corms, attacks leaf bases within the sheathing outer base, and finally causes collapse of the plant with typical neck rot (an alternative name for the disease). Spread of infection can be less dramatic in dry conditions, and a flower spike may be produced. The disease can persist in the soil or be carried on corms, which exude a gum which hardens to a characteristic lacquer-like finish, useful in identifying infected corms. Control involves crop rotation, roguing, the dipping of cleaned corms, and their dusting at planting. This pathogen also affects crocus. Unusually, the bacterial blight of gladioli, caused by a *Xanthomonas* species, is not known to occur in the UK although it is a recognized problem in the USA and several European countries.

Fungal Diseases

Fusarium diseases are very important for all 'bulb' crops, causing heavy losses annually. Their control has proved to be the most intractable problem of modern 'bulb' growing, and is given high priority in many research programmes. Basal rot of narcissus is caused by *Fusarium oxysporum* f. sp. *narcissi*. At lifting, infected bulbs are soft, especially near the base, discoloured a reddish brown, and frequently rootless. The rot progresses rapidly, especially in warm (above 18°C) conditions, the whole bulb becomes soft with white or pink mycelial wefts like cotton wool. Eventually the bulb dries, shrinks and becomes hard, brittle and mummified (Fig. 9.2). The fungus persists in the soil, and has been isolated from soils that have never grown narcissi. Infection occurs in the field through old, dying roots in summer, and in store and during planting from spores released from mycelium on infected bulbs. The incidence of the disease can be reduced by delaying planting until soil temperatures fall in the autumn, by dipping bulbs in fungicidal solution for 30 min as soon after lifting as possible, and by routine h.w.t. as soon as possible after lifting. Whilst an annual lifting of narcissus bulbs and h.w.t. gives better disease control than the usual biennial lifting, this is not considered a practical proposition in the UK, although general practice in The Netherlands. The breeding and selection of resistant cultivars is a long-term goal, and some resistant cultivars have already been identified, e.g. St Keverne.

Another less well understood disease of narcissi is neck rot. It is increasingly becoming more important. Commonly the fungus extractable is *Fusarium oxysporum*, but in some cases other species seem to be responsible, and the symptoms vary from a dry ginger coloured necrosis to a chocolate coloured wet rot. Infection starts at the neck of the bulb and the fungus grows down the old flower stalk to infect the bulb. The disease is spread during storage and handling, when aerial spores are disseminated. Presently recommended control measures are as for basal rot.

Fusarium bulb rot, also called 'sour' (*zuur* in Dutch) from the smell of the

Fig. 9.2. Narcissus bulb infected with *Fusarium oxysporum* f.sp. *narcissi*. Note the general brown discoloration and, centre, fluffy white mycelium, often tinged pink. (Photograph courtesy of *Horticulture Research International*, the copyright holder.)

infected bulbs, is an important disease of tulips caused by *F. oxysporum* f. sp. *tulipae* which came into prominence with the increased popularity of the susceptible Darwin hybrid forms. On newly lifted bulbs, pale brown depressed flecks on the outer fleshy scale coalesce slowly to form large brown patches. When these reach the base plate, further progress is rapid, often associated with pink to mauve mycelium bearing spores. Infected bulbs often produce copious gum (up to 2 ml per bulb) which is initially a sticky clear liquid which later hardens and darkens, and also smells of pear drops, but later this changes to the characteristic sour smell (Fig. 9.3). The ethylene emitted by

Fig. 9.3. Tulip bulb showing gum exudation at the bulb tip and equatorially. (Photograph courtesy of *Horticulture Research International*, the copyright holder.)

infected bulbs has physiological effects on healthy bulbs within the store, such as inducing them to produce gum and damaging the developing flowers (see later in this chapter).

In the field, infected plants can often be identified from the early senescence but upright posture of the shoots which often turn purple, and the fungus persists in the soil for many years, even in the absence of host plants. Control measures are based on early lifting before the tunic has become papery (at this stage the bulbs are resistant to infection because of the presence of a lactone), on dipping bulbs in systemic fungicides immediately after lifting, on planting the bulbs in autumn when soil temperatures are below *c.* 10°C, and on the use of pre-planting dips of fungicide. Because the fungus persists in the soil, the longer the time interval between successive crops, the better. Screening of existing cultivars and various crosses between resistant and susceptible ones have indicated a simple inheritance of resistance, even from one resistant parent. This has led to a programme of work at the Institute for Horticultural Plant Breeding in The Netherlands aimed at producing resistant parents for use by private breeders in the development of finished varieties.

Because the fusaria are ubiquitous, with a wide host range, serious diseases similar to the above affect many other 'bulb' plants, including gladiolus (fusarium yellows), freesia (fusarium corm rot), iris (fusarium basal rot), lily (fusarium rot) and minor bulbs including crocus, crocosmia, ixia and sparaxis.

A number of serious diseases of many 'bulb' species are a result of attack by fungi of the genus *Botrytis*. The diseases are generally commonest at low temperatures and at high humidity, and spread is encouraged by rain, water splash and persistent water films on the plants. These diseases are therefore most common in the field, where these environmental factors cannot be controlled. In glasshouses, attention to plant spacing, temperature, ventilation and care with watering are simple and effective means of minimizing the incidence of botrytis diseases.

Probably the best known of this group is tulip fire caused by *B. tulipae*. Called *vuur* in The Netherlands and 'tulip mould' in the USA, it occurs wherever tulips are grown. The life history of the causal organism is complex, but can be divided into above-ground and below-ground phases (Price, 1970). In the former, the first indications in the field and in forced crops are infected shoots whose leaves do not open but form sclerotia and spores (Fig. 9.4). In the vicinity of such shoots, called 'primaries' (*stekers* in Dutch), patches of severe leaf spotting occur as a result of spore dispersal and germination. The lesions often remain small and non-aggressive, but in damp conditions they enlarge and more spores are produced, leading to considerable loss of leaf area, and, later, of flowers. Even slight spotting of flowers is sufficiently unsightly to render them unsaleable. Below ground there are few indications of future problems. From infected mother bulbs, the pathogen infects daughter bulbs, providing an underground transfer of the disease between successive host generations. Daughter bulbs may bear small black sclerotia on or under the tunic, or, if only lightly infected, slight lesions on the scales. After planting the emergent shoot becomes infected and becomes a primary. Sclerotia can also become detached from the old mother bulb and remain in the soil as a persistent source of further infection. Several measures can reduce disease incidence. Removal of bulbs with many sclerotia at bulb cleaning time, dipping bulbs in systemic fungicide between lifting and replanting, foliar sprays in the field, and the removal of primaries whenever practicable all contribute to minimizing disease incidence. Timely deheading of flowers in the field or garden is also a valuable aid to control because abscinded perianth parts, especially if adhering to a leaf with a film of water, provide an excellent substrate for spore germination and fungal growth. A similar disease affects hyacinths. It is also called fire, and the causal organism, *B. hyacinthi*, also attacks lily and muscari.

Gladioli are attacked by *Sclerotinia draytoni*, causing the disease botrytis

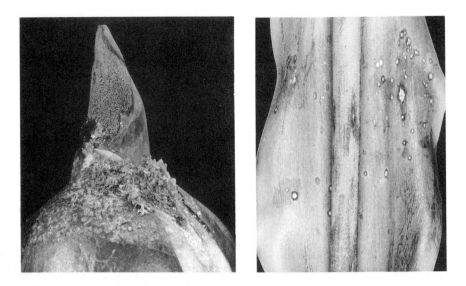

Fig. 9.4. Left, 'primary' tulip bulb infected with *Botrytis tulipae*, the causal organism of fire. These infections release large quantities of spores which affect neighbouring plants. Right, symptoms produced on a leaf by germinating spores, the mycelium can die under dry conditions leaving non-aggressive lesions, or can spread aggressively to produce rotting. Similar lesions occur on flowers, making them unsaleable. (Photographs courtesy of *Horticulture Research International*, the copyright holder.)

rot (also known as spongy rot, core rot, corm rot and petal spot). Like tulip fire, there is a corm infection which is transferred to above-ground parts, which turn yellow then brown, sporulating profusely. These spores produce aggressive or non-aggressive lesions on foliage and flowers, and spots can develop on apparently clean flowers whilst in transit to market. There is a transfer of infection from mother to daughter corm below ground. Control measures are as for tulips, with the addition of h.w.t. of the corms, after an initial soaking period, at 53°C for 30 min, followed by rapid drying and low-temperature storage before replanting.

The most serious and widespread diseases of gladiolus wherever this crop is grown is dry rot. The pathogen, *Stromatinia gladioli*, causes the well-known yellowing of leaves about 6 weeks after emergence; this is soon followed by browning and the rapid death of the plant. Plants from cormels die more rapidly, but late-infected flowering size plants can escape death, despite yellowing and wilting, and the loss of many roots. Small brown lesions on the corms develop into a dry rot during winter storage, involving the whole corm and transforming it into a shrivelled, hard mummy. Affected growing plants cannot be treated, but corms can be given h.w.t. as recommended for botrytis rot above. Soil sterilization is also effective. The disease also affects crocus, freesia, snowdrop, narcissus and others.

Fire and smoulder are two diseases affecting narcissi caused by *Sclerotinia polyblastis* and *S. narcissicola*, respectively. The former seriously affects flowers, especially of tazetta cultivars, but is not so important with flower crops picked early for market. The fungus overwinters on plant debris and in the soil, and can be controlled by good hygiene and timely sprays. Smoulder occasionally reaches epidemic proportions, causing leaf and flower spotting. Small black sclerotia are present on bulbs, and these germinate in spring to produce ascospores. The infected emerging shoots also release spores, and the fungus grows down into the neck of the bulb and forms more sclerotia. In moist conditions, spread of infection in the field is rapid. Spraying can reduce disease spread in the above-ground phase. Both fire and smoulder are controlled by h.w.t.; in particular, it is unusual for smoulder to occur in the first year after planting.

Snowdrops seem less affected by fungi and bacteria than other 'bulb' plants, but are affected by grey mould (*Botrytis galanthina*). Emerging infected shoots are grey, completely enveloped in sporulating mycelium, and resemble small mushrooms. The leaves rot, and this soon affects the bulb which collapses to a pulp. Infection of healthy bulbs is believed to occur mainly by the underground route in contaminated soil or from introduced infected bulbs. Control includes avoiding introducing affected stock, not growing snowdrops in soil known to be affected, sterilizing affected soil, and h.w.t. at 42°C for 1 h in water containing fungicide.

Botrytis diseases of alliums, anemone, chionodoxa, colchicum, dahlia, iris (both bulbous and rhizomatous), ixia, lily, lily-of-the-valley and muscari have also been reported.

Penicillium is a genus of fungi usually associated with 'bulbs' in storage. Some species, at least in some circumstances, appear to be saprophytic on dead outer tissues, but can be parasitic given appropriate environmental conditions. Other species are clearly pathogenic, but may require some damage to the host tissue to become active.

The most aggressive and important *Penicillium* disease is bulb rot of bulbous irises caused by *P. corymbiferum*. Although entry of the fungus is mainly through wounds, healthy bulbs can be attacked in the field. The rotting and softening of the bulbs occurs mainly during storage from an initial large lesion usually on the side of the bulb, on the outer fleshy scale, but concealed by the tunics. Blue-green spore masses develop on the site of the original lesion, and the rot spreads to other scales via the base plate. The bulb becomes wet and collapses. Bulbs infected late, or only lightly, survive to produce a flower. Prevention of damage to bulbs at lifting is important, as is a few days at 17°C after lifting because this 'seals' existing damaged areas from infection. Prevention of root emergence in store also reduces infection, as does a pre-planting fungicidal dip.

Storage rot of damaged corms of gladioli is caused by *Penicillium gladioli*. A few large, sunken, red-brown ill-defined lesions beneath the corm scales cover deep cones of decay which also contain some sclerotia. The rot enlarges rapidly in low temperatures and high humidity. The pathogen also attacks storage organs of crocosmia, crocus, freesia, scilla and tigridia. Because fungal entry is through wounds, care in handling corms is important, and warm storage (29°C) immediately after lifting slows fungal growth. Dusting or dipping the cleaned corms in an appropriate fungicide provides additional protection.

Lily bulbs in store can be affected by two species, *P. brevicompactum* and *P. corymbiferum*, following invasion through damaged scales. The latter species affects tulips, occasionally causing rots under somewhat ill-defined circumstances, and appears also to be associated with a poorly understood condition called 'chalking', a physiological disorder which makes the bulbs dry, opaque and chalk-like.

Because of the general requirement for wound penetration, some *Penicillium* infections are associated with rapid propagation methods that involve cutting up of bulbs. Hyacinth bulbs, when cut or scooped to induce many adventitious daughter bulbs encourage attack by weakly pathogenic fungi, including *P. hirsutum*, *P. cyclopium* and *P. corymbiferum*. In propagating narcissus bulbs, also, care must be taken to incubate twin-scales with fungicide effective against penicillia to prevent losses of bulbils.

Two leaf diseases of narcissus grown outdoors in the Southwest are leaf scorch (*Stagonospora curtisii*) and white mould (*Ramularia vallisumbrosae*); within the UK the latter is largely confined to this area. The life history of the leaf scorch fungus resembles that causing smoulder: leaves are infected from the neck of the bulb at emergence, producing tip lesions which are a source of secondary spread to other leaves and flowers. White mould first appears as pale sunken spots or streaks, which coalesce in wet conditions to affect large areas of leaf, whose undersides are covered with white, powdery masses of spores. Late flowering cultivars are worse affected, and within a crop the disease incidence worsens year by year. Timely sprays with appropriate fungicides are recommended, although near flowering time these can leave unsightly deposits which lower flower quality.

The most common and serious leaf disease of both bulbous and rhizomatous irises worldwide, is leaf spot. The causal organism is *Mycosphaerella macrospora*, and leaves and flowers (but not the underground parts) are affected. Disfiguring spots make flowers unsightly and unsaleable, whilst severe leaf infection leads to early leaf death, low yields and consequently poor flowering in subsequent seasons. During the growing period, short-lived spores are dispersed by wind and water splash and generate secondary infections. Overwintering of the fungus is mainly as mycelium in the old leaf bases, which sporulates in spring. Spraying before symptom appearance is recommended in areas where the disease is prevalent.

Ink disease of irises is named from the black, ink-like, crusty streaks or patches on the outer tunics of the bulbs; it is a serious leaf disease of both bulbous and rhizomatous irises. The fungus responsible is *Drechslera iridis*. Severely affected bulbs rot away completely soon after planting, leaving a shell containing a little black powder. If less serious, aerial growth can be stunted, but infected daughter bulbs can still be produced. When planted, these can infect healthy neighbours, whose leaves then develop the early symptoms of black streaks which can progress to early leaf senescence with a typical red-brown colour. Spores produced from affected leaves are dispersed by wind and rain splash, and set up secondary leaf lesions. Regular spraying controls the leaf spotting phase, but bulb dips have not proved very effective. General hygiene in the field, especially using the 'burning-over' technique for earlier and improved flowering of bulbous iris, allied to leaf spraying, are the best control measures available.

Anemones suffer from attack by two mildews, downy mildew caused by *Perenospora* and *Plasmopara* spp., which cause leaf curl and plant stunting, and, occasionally, powdery mildew (*Erysiphe ranunculi*). Control of the downy mildews is difficult. Foliar sprays are not economic, especially in wet, windy conditions, and the resting spores remain viable in the soil for several seasons.

Some diseases of 'bulbs' which occur in the field are also troublesome on forced 'bulbs'. Where the environmental conditions are sufficiently different (higher temperatures, lower humidity), growth of the crop is more rapid and the plants are discarded after flowering, so that many problems are avoided. However, the higher temperatures of the growing medium can favour the activity of some pathogens, such as those causing tulip fire (*B. tulipae*), fusarium bulb rot (*F. oxysporum* f.sp. *tulipae*), root and soft rot (*Pythium* spp.), grey bulb rot (*Sclerotium tuliparum*), and shanking (*Phytophthora cryptogea*) of tulip. Some of the same or related fungi also attack lilies and other 'bulb' plants. Regular inspection and the removal of infected plants can be practised in the forcing house, the use of disease-free or sterilized growing media is a sensible precaution, and the dusting or dipping of bulbs at boxing is a sound prophylactic measure.

A comparatively recent disease of anemone, first discovered in the UK on corms imported in 1978, is leaf curl, caused by *Colletotrichum* spp. Little is known about the control of this disease, but care in importing or introducing new material onto a holding which has no previous history of the disease is essential.

Although the fungal diseases listed above seem numerous, there are many others which are not described in detail, partly because they have been the subject of less research than the important ones. They are of minor interest because they occur in very few places, with unusual climates, or occur only occasionally, affect only minor crops, or cause only little damage.

Fig. 9.5. Healthy (left) and 'broken' (right) flowers of tulip, a result of infection with tulip breaking virus. (Photograph courtesy of Dr A.A. Brunt.)

To the individual specialist grower, they can be serious, but in the overall picture of 'bulb' growing they are of less interest. Detailed information is available in *Moore's Diseases of Bulbs* (Brunt *et al.*, 1979).

Virus Diseases

Paradoxically, some effects of virus infection have been known for a very long time, but the cause has been recognized only comparatively recently. Tulip 'breaking', which produces the bizarre and much prized coloured/white patterns on tulip perianths was thought to be due to a natural, but sometimes slow, development of plants with plain coloured flowers which were called 'mother' or 'breeder' tulips (Fig. 9.5). The rich merchants in The Netherlands became obsessed with possessing broken tulips, particularly those with violet or crimson stripes on a pure white ground called *bijbloemen* and *roses*, respectively. The associated financial speculation called 'tulipomania' culminated in the peak prices (equivalent to hundreds of pounds for a single bulb) of 1636, followed by the crash early in the following year. However, the price of a single bulb of a well 'broken' tulip in the UK was £100–£150 even in the decades prior to 1850. The first ideas that the condition was a result of virus

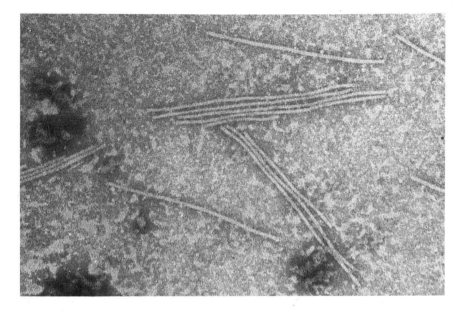

Fig. 9.6. Electron micrograph showing the flexuous filamentous particles of narcissus yellow stripe virus (× 75 000). (Photograph courtesy of Dr A.A. Brunt.)

infection were not published until 1919, sap transmission was demonstrated in 1927, and the vectors of tulip breaking virus were identified as aphids in 1928. Since then modern methods of serology and electron microscopy have demonstrated that all bulb plants can be infected by viruses, and there are many viruses, which differ greatly in morphology, specificity, severity, symptom expression and mode of transmission. In the past 20 years or so, the viruses and their diseases in 'bulb' plants have been extensively studied, most have been described and characterized, and rapid methods of identification have been developed. For instance at least 13 viruses are known to infect tulip, with 15 confirmed in narcissus.

It is curious that the symptoms caused by the tulip breaking virus, originally regarded as a desirable feature in tulips, were in reality those of a damaging disease; the severe strain of the tulip breaking virus, particularly, causes stunting of aerial growth and reduces the yields of daughter bulbs. The resulting small bulbs are often discarded by growers, so the disease is partially self-limiting in the commercial situation. In narcissus, symptoms of virus diseases range from the severe chlorosis and distortion of leaves and scapes by narcissus yellow stripe virus (Fig. 9.6) to the inconspicuous and easily overlooked pale leaf markings caused by narcissus latent virus. Short-term losses of yield resulting from virus infection are difficult to demonstrate, although over long periods, yield losses in individual stocks had led to ideas of stock 'degeneration' even before the viral cause was identified. Some viruses seem to cause little or no damage or yield loss, but even in these cases there are potential dangers. If a virus were transmitted to another cultivar or another species, there could be serious consequences for flower quality or 'bulb' yield. It might also cause losses if it became a co-infector with another mild virus.

Although the most severe diseases can be recognized with some certainty from the symptoms, other methods are required to determine the causal virus. Some viruses have a broad host range; cucumber mosaic virus, for example, infects many crop and weed plants which act as a reservoir for aphid trans-mission to crocus, gladiolus, iris, lily, narcissus, tulip, and other 'bulbs'. Others are highly specific, and there are several viruses for which narcissus is the only known host. Some cultivars display more severe symptoms than others, infection of an individual plant with more than one virus commonly occurs, with confusion of symptom expression, and there are many virus diseases of 'bulb' plants, so the situation is complex. To identify viruses accurately, it is necessary to examine some infective sap using an electron microscope to determine particle shape and size. Serological methods can also be used to detect and identify individal viruses using specific antisera. In this way viruses can be accurately identified.

Several characteristics of virus diseases demarcate them from other diseases where physical or chemical control measures are directed against the pathogen, but allow the host to survive. In general, methods of controlling virus diseases aim to restrict spread. Virus diseases are systemic, all parts of the plant are usually infected, with two important exceptions. Seed is usually free of virus particles, although a few, like the nematode borne ringspot viruses, are seed borne. However, this feature can be exploited only if the plant can readily be reproduced true-to-type from seed in a reasonable time – a situation which does not hold for the majority of 'bulb' plants, which are vegetatively propagated clones, frequently sterile, and often take several years to grow to reproductive maturity. However, if care is taken to isolate the seedlings in a breeding programme from contact with virus vectors and there is no mechanical transmission, plants can be kept free of virus disease until released into field conditions.

The second exception is the apical meristems of the plant. Careful and sterile removal of the apical dome with as little associated tissue (leaf primor-dia) as possible, compatible with the survival of the explant, provides a source of virus-free material (but this must be tested). Transfer of meristem tips to sterile artificial media and propagating them *in vitro* is now routinely used to produce virus-free material of many plants, the technique being called meris-tem-tip culture. The disadvantages of small size, difficulties of transfer of plantlets to growing media and the expense of the operation are largely outweighed by the high rate of multiplication achievable.

These methods, allied to the roguing of obviously infected plants, form the basis of control measures. It is important to do the roguing when symptoms are most obvious, before aphid numbers increase appreciably, and before the soil temperature rise of spring increases soil nematode populations. Control of vectors is possible in some situations, such as nematicidal soil treatments and the use of aphid-proof screens to enclose areas used for propagating and growing-on seedlings and young plants prior to their transfer to the field. Viruses are carried in aphid stylets, so that transmission takes only minutes, and can occur without host colonization. Systemic insecticidal sprays are of little value in this situation and might even increase the rate of spread because the chemical irritates the aphids into increased feeding activity before death. The use of protective oil sprays against the aphid spread of virus diseases has been developed in The Netherlands, especially in lily production, but results in some loss of bulb yield, presumably by interfering with gas exchange through the stomata. The isolation of newly bought-in stocks and healthy 'mother blocks' in the field, and the use of barrier crops, are widely used techniques to prevent the rapid re-infection of 'clean' stocks in the field.

The selection of the most vigorous, healthy looking plants and propagating these (as 'greenstocks') in isolation from others, was a reasonably successful technique for maintaining stock health and freedom from virus diseases that has now largely been superseded by the production of tested, virus-free stocks by meristem-tip culture. 'Bulbs' that are virus-free, or virus-tested (a preferred term, indicating freedom from all *known* viruses) are now available for many cultivars of gladiolus, iris, lily and narcissus.

Viruses affecting narcissus are listed below, with their vectors.

1. Aphid transmitted:
i)   narcissus yellow stripe (NYSV);
ii)   narcissus white streak (NWSV);
iii) narcissus latent (NLV);
iv)   narcissus degeneration (NDV);
v)   cucumber mosaic (CMV);
vi)   broadbean wilt (BBWV).

2. Nematode transmitted:
i)   arabis mosaic (AMV);
ii)   strawberry latent ringspot (SLRV);
iii) tomato black ring (TBRV);
iv)   raspberry ringspot (RRV);
v)   tobacco ringspot (TobRSV);
vi)   tomato ringspot (TomRSV);
vii) tobacco rattle (TRV).

3. Vector unknown:
i)   narcissus mosaic (NMV);
ii)   narcissus tip necrosis (NTNV).

Gladioli are affected by nine viruses.

1. Aphid transmitted:
i)   bean yellow mosaic (BYMV);
ii)   pea mosaic (PMV);
iii) bean common mosaic (BCMV);
iv)   cucumber mosaic (CMV);
v)   soybean mosaic (SBMV);
vi)   gladiolus latent (GLV).

2. Nematode transmitted:
i)   tobacco rattle (TRV);
ii)   tobacco ringspot (TobRSV);
iii) tomato ringspot (TomRSV).

Lilies are affected by seven viruses.

1. Aphid transmitted:
i)   tulip breaking (TBV);
ii)   lily symptomless (LSV);
iii) cucumber mosaic (CMV);
iv)   lily rosette (LRV).

2. Nematode transmitted:
i)   tobacco ringspot (TobRSV);
ii)   tobacco rattle (TRV);
iii) arabis mosaic (AMV).

The above lists for three species give an idea how many viruses can affect a 'bulb' species and the numbers common to all three or two out of three, despite their hosts being members of separate families. In the minor 'bulbs' there are fewer records of viruses, probably a reflection of the lack of research done on them rather than a true indication that they have fewer virus diseases.

Schemes for the production of virus-tested material require the setting up of national organizations, such as the Nuclear Stock Associations, under the aegis of the National Farmers Unions of England and Wales, and, separately, of Scotland. The starting points are a few meristem-tip cultures of selected cultivars, produced and virus-tested by a research organization or a commercial micropropagation company with trained staff, the necessary electron microscope facilities, and the capacity for detecting and identifying viruses by immunodiffusion serological tests, enzyme linked immunosorbent assay and/ or immunosorbent electron microscopy. The explants grow into plantlets which can be multiplied *in vitro* to manageable numbers, before transfer to grow in sterilized soil or growing substrates. When their 'bulbs' are sufficiently large, further propagation can be done, usually using twin-scaling or chipping, ending with a small stock of plants. It is necessary at all stages to continue testing for the presence of viruses in case attenuated virus infection has passed earlier screening, or some reinfection has occurred despite all safeguards. The detailed procedures used for bulking up vary with species, but in general the rapid multiplication *in vitro* is expensive in facilities and labour, so that transfer to soil/growing media (a difficult operation, involving some losses) is done sufficiently early to allow further care using less sophisticated equipment and techniques. There is a conflict between the early rapid multiplication of plantlets and the need to grow these to produce easily handled bulbs, i.e. between plant numbers and total weight, and the process is managed so as to achieve greatest efficiency, measured as the cost of each individual plant at release to the grower. It can take up to 11 years from establishing a meristem-tip culture to the date of release of material to commercial producers for further propagation and sale.

It is usual for further bulking of the stock to be done by licensed growers in isolation from other stocks, and taking all reasonable precautions to prevent reinfection. When sufficient stocks have been produced, they are distributed to the participating growers for commercial use. Stocks in the field do become re-infected, at rates depending on many factors, but after some years it usually pays to discard old stocks and replace them with new virus-tested ones. Commercial stocks of the narcissus cv. Ice Follies, first registered in 1953, were still found to be mainly virus free 34 years later, and individual plants of several narcissus cultivars of the same era have been shown to be free of viruses. In contrast, surveys in the mid-1970s failed to find any virus-free bulbous iris or lily plants. It is clearly necessary that nuclear stock schemes are continuously supplied with virus-tested plants from the initial material maintained *in vitro*. New cultivars are introduced into such schemes as demand for them increases, but the volume of stock being dealt with, and the large number of species and cultivars preclude all cultivars being available as virus-tested stocks. By 1984 over 70 species and cultivars had been freed of virus, as individual plants, but not necessarily as stocks.

Freedom from virus diseases is a vital component of crop health, which contributes to the viability of the industry and sustains exports.

PHYSIOLOGICAL DISORDERS

A number of disorders of 'bulb' plants exist for which there is no known causal pathogen. Others are known to result from exposure to adverse environmental conditions, either during storage or during the growing period, which do not allow normal growth of the plant at that time. In many cases the cause is unknown, but pre-disposing factors have been identified, with experience, as being associated with the disorder. In others, with well-defined symptoms, causes are well established and remedial measures are available.

Frequently a given symptom (such as flower death) can have several causes, whose disentanglement can be difficult because the problem might have occurred much earlier, perhaps during 'bulb' storage, but its effects are not apparent until the normal time of flowering. It is often important to be able to establish in such cases when the damage was done, especially if the 'bulbs' were bought-in by the grower, and he wishes to be recompensed by the supplier for the loss of the flower crop. Whilst it is possible by dissection to determine from the aborted flower bud size an indication of the time of death, the time-scale of several months frequently makes this too imprecise an estimate to allow blame to be apportioned.

Among the environmental factors that can lead to non-parasitic diseases are inappropriate temperatures for the phase of development (during both storage and growing), the gaseous environment, exposure to chemicals such as herbicides, mechanical damage by hail, and effects of waterlogging. The alert grower will be aware of many of these hazards. Some environmental effects can be quite subtle, such as early lifting and storage in a dry atmosphere which can produce dry hard 'bulb' bases which restrict root penetration and force them to grow upwards inside the outer scales of narcissus and tulip bulbs. With good watering, such forced bulbs perform adequately, but flowers can die if the soil become dry. Occasionally, physiological disorders result from the indirect effects of pathogen attack. One of the best known is the damage to healthy tulip bulbs during storage by ethylene generated by *Fusarium*-infected bulbs.

Flowering failures occur with many 'bulbs' grown in the field. Some of these are direct effects of overcrowding, where the natural increase in daughter 'bulb' numbers leads annually to smaller and smaller propagules, which eventually fall below the critical size for flowering. Insect or fungal attack, or too-severe or badly timed h.w.t. can kill growing points within 'bulbs' and lead to 'grassiness', as the loss of apical dominance stimulates the outgrowth of adventitious shoots, as seen in the scooping, cross-cutting and chipping techniques used for artificial propagation. When lifted, there are many small 'bulbs'.

Pre-emergence flower failure is common, and often points to inappropriate temperature treatment. In the h.w.t. of narcissi, following the correct procedures results in over 90% of flowers being marketable. If not, damage to all parts may result. Roots may be killed by late treatment, so the planted bulbs in the field pass a whole season with no roots and little growth, although the plants usually survive. Leaf tips can be damaged, and flowers split, distorted, dwarfed or killed. Much of this damage can be avoided by warm storage of the bulbs before h.w.t., care with the temperature and duration of treatment, and treating at the optimum time, as described earlier. The large tulip bulbs sold for outdoor flower production in parks and gardens, and those used for forcing, nearly all produce flowers, but about 1% fail to do so, and have a single large leaf like that produced by a bulb below flowering size. In a rare condition of tulip called 'antholyse', the leaf complement is normal, but the shoot tip ends in a thong-like process. The reasons for both these failures of flower initiation are not known.

Much more common are flower death after initiation, called blindness, blasting (more usually applied to flower drying just before anthesis) or abscission (in lilies, where the dried flower falls off). In tulip, the time of flower death determines the size of the dried-up flower, and also the resultant stem extension (Fig. 9.7). In Chapter 5, the effect of the flower, and especially the ovary, in promoting stem extension was described. If the flower is affected at an early stage, there is little stem extension, and the flower bud appears as a small, papery, pale brown structure above the dried-up last internode of the shoot. In this situation, leaves appear normal but may be smaller than usual. A dried tulip flower bud shorter than 1 cm indicates that it died before the bulb was planted. Damage late in development produces only slight symptoms such as poorly developed anthers, straw coloured in cultivars which normally have black ones, white tips to perianth parts, failure of degreening of the coloured parts of the flower, and lobed or notched (rather than smooth and entire) perianth parts (Fig. 9.8).

There are several causes of blasting as can be observed from the different incidences in neighbouring boxes during forcing, suggesting a tendency to blasting which can be caused by slight variations in watering, in glasshouse temperature, in the depth of growing medium, or by bulb factors such as size and the extent of rooting (perhaps associated with hard base). Other factors that have been implicated include forcing small bulbs very early at high temperatures and starting the low-temperature treatment very early. There are also suspicions that ethylene generated from *Fusarium*-infected bulbs can blast neighbouring healthy bulbs in the same forcing box. Experimental work indicates that with adequately sized bulbs, higher forcing temperatures are better than a low one (15°C), with 17°C the optimum, and that nitrogen fertilizer applied at housing decreases the incidence of blasting.

Blasting in tulip must be distinguished from 'bud necrosis' (*kernrot*) which is a black/brown wet rot of the flower and often the top internode which starts when the bulb is in store, with no external indication of anything untoward (Fig. 9.9). Research in The Netherlands has indicated that this disorder is complex; and is initiated by ethylene generation in store from *Fusarium*-infected bulbs. The gas (which is biologically active at concentrations as low as 0.08 v.p.m., and can reach several v.p.m. in poorly run bulb stores) deforms the flower bud by delaying elongation of the perianth parts, producing an 'open' bud with protuberant anthers which are attacked by *Rhizoglyphus* and *Tyrophagus* mites. Fungal and bacterial spores carried in by the mites then colonize the anthers and set up the rot. In the absence of mite infestation, there is no necrosis and the plants are either normal, or have flowers only slightly reduced in size.

A third disorder of tulips, called 'heating in transit' because it commonly occurred in bulbs being exported and accidentally subjected to high temperatures on the quayside, results in the stem tip and flower being deformed

Fig. 9.7. Tulip flowering disorders. From the left: (1) normal flower, (2) blind flower, affected late, as shown by normal stem extension until the last internode. Flower white and dry, (3) blind flower, affected early. The flower is the small white central object, stem extension severely restricted, (4) shoot affected as well as flower, only one leaf remains, and (5) shoot completely killed. With increased damage, there is a production of green aerial extensions to the daughter bulb scales of plants (4) and (5). (Photograph courtesy of *Horticulture Research International*, the copyright holder.)

Fig. 9.8. Flower of tulip cv. Apeldoorn showing symptoms of late blasting. Note white, dried-out tips to perianth parts, pale shrivelled anthers (normally black) and poorly developed gynoecium. Similar symptoms result from exposure to ethylene. (Photograph courtesy of *Horticulture Research International*, the copyright holder.)

into a green knob-like structure. A commercial treatment based on this effect (*blindstoken*) was described briefly in Chapter 7.

Easter lily flowers also suffer from bud blasting or floral abortion, usually of tertiary flowers in the inflorescence. The base of the flower bud at any length between 2 and 50 mm becomes light green, then yellow, shrivels and finally turns brown but does not absciss. A series of investigations suggested the following causes: high flower number per inflorescence, sudden moisture deficits, high soil nitrate status, low light. For mid-century lilies, a distinction has been drawn between blasting and bud abscission, both of which are affected by light and by ethylene. The latter is more critically related to the developmental stage of meiosis in the anthers. It is now current practice to decrease the incidence of both disorders, especially in winter, by increased

Fig. 9.9. Left, flower blasting of tulip at an early stage. The flower is small and white and is enclosed in the third leaf, above the dead, unextended top internode. Right, bud necrosis, showing the black wet rot which has affected the whole shoot except the lowest leaf. (Photograph courtesy of *Horticulture Research International*, the copyright holder.)

spacing and/or the use of supplementary lighting. Recent work in the USA on flower buds of *Lilium elegans* showed a marked decrease in respiration, accompanied by peaks of evolution of ethylene and CO_2, and a fall in sucrose content accompanied by growth cessation, prior to abscission. None of these effects was observed in buds developing normally. These observations indicate that the depletion of carbohydrate reserves essential for growth and development is responsible for stopping the growth of the buds which then absciss.

Like tulips, Dutch iris flowers suffer from failure of flower initiation (called 'three leaf', *driebladers*, because non-flowering plants have three leaves) and from flower blasting. A higher percentage of bulbs near the critical size for flowering can be induced to flower by high-temperature treatment shortly after lifting, or by ethylene treatment. Blasting is most common in the low irradiance conditions of winter forcing. The flower buds and their enclosing spathe leaves turn brown and become papery and dry, but do not absciss. The major cause appears to be preferential distribution of carbohydrates to the daughter bulbs rather than to the flower. In better light, there is more available assimilate so the flowers do not blast. Increasing the sink strength of the flower with cytokinin or gibberellin also helps prevent blasting in iris, and also in lily.

Some plants require well-defined temperature regimes to flower. If these are not provided, the percentage of flowering in a population is much reduced or completely suppressed. Such behaviour may vary considerably within a genus. The flowering of *Nerine bowdenii* is unaffected by temperature within the range 9–25°C during the growing period, but effects appear the following season, with 83–99% flowering up to 21°C, but only 39% at 25°C. Flower differentiation is not completed at the high temperature and the inflorescence aborts without emerging from the bulb. Experimental work showed that, in contrast, almost all *N. sarniensis* bulbs flowered at 22°C, compared with only a third at 14°C. *N. flexuosa* requires growing temperatures within the range 9–17°C, and with a combination of growing at 9 and 21°C storage all bulbs flower. *Clivia miniata*, a popular evergreen pot plant, has a cold requirement of 10°C for 60 days. Without this there are no flowers, despite normal vegetative growth.

An unmistakeable condition affecting many 'bulb' flowers is that called 'topple'. It involves a collapse of the stem at about the time of anthesis or even after picking so that the flower hangs down (Fig. 9.10). Best known in tulips, it also affects iris, lily and gladiolus. In tulip, the first symptom is a watery appearance of the part of the stem which is at that time extending the most rapidly, usually just below the flower or the top leaf. This is soon followed by the inward collapse of these tissues, and the toppling of the flower. It affects forced flowers; outdoor ones hardly ever. It has long been associated with calcium deficiency, and by conditions which reduce transpiration (like high atmospheric humidity), as calcium is translocated in the transpiration stream. Surprisingly, a low calcium content of the bulbs themselves is no guide to predisposition of a plant to topple, despite the subnormal calcium concentration of that part of the stem which collapses. A high plant nitrogen content does seem to be beneficial. A poor root system can restrict water uptake, and energy saving methods which reduce glasshouse ventilation and increase humidity both make matters worse. A reduced glasshouse temperature can decrease the incidence of topple, but care is needed to ensure that it does not raise the humidity, which should be around 80% r.h. When the condition is first observed in the glasshouse, spraying with dilute calcium nitrate solution is said to reduce losses.

There are two forms of 'loose bud' of hyacinth, one of which seems to resemble tulip topple, within the constraints of differing morphology. In both, the inflorescence becomes detached at its base and carried upwards (hence *spouwen* or *cracher* in Dutch). Physiological loose bud involves a sap infiltration

Fig. 9.10. Tulip topple, showing collapse of the stem in the top internode region. One flower has had its leaves removed to show the restricted length of affected stem. (Photograph courtesy of *Horticulture Research International*, the copyright holder.)

of the base plate, producing a water-soaked appearance and a narrow longitudinal cavity in the peduncle from which radial fissures develop into cavities. As the leaves grow, they exert an upward pressure on the inflorescence, and the weakened scape fractures, leaving a short stump in the leaf cup. Because this is sap filled, there is initially no loss of turgor by the inflorescence; as soon as contact with this liquid is lost, the inflorescence becomes flaccid and collapses. The second form is believed to be entirely mechanical, as there are no splits, cavities or water soaking of the scape. The condition occurs most frequently following rapid increases in temperature which encourage faster growth.

Occasionally floral malformations occur, such as the 'bullhead' condition of double narcissus cultivars like Cheerfulness. Up to 10% of flowers can be affected; the flowers fail to emerge from the spathe, which remains dry and membranous, giving a drumstick appearance. The normal double flower is a result of petaloid anthers, but the bullhead has extra perianth parts which appear to hinder the extension of the corolla tube so the spathe is not split. Marking affected plants has shown that they produce similar abnormal flowers in the next season, indicating either a virus or a genetic cause. Abnormal flower part numbers occur in other species, but only as occasional aberrations of little commercial consequence. An extra member of each perianth whorl giving eight-'petaled' tulips makes for a more attractive flower than the loss of one member, giving a cruciform four-'petaled' flower. Extremes of floral part numbers can follow both higher and lower temperature storage, depending on cultivar, but in tulip, lower temperatures tend to result in increased part numbers, perhaps a result of the larger apical meristem produced. Bulbous iris flowers with aberrantly increased floral part numbers result from temperatures below zero during flower differentiation. Quaternate, quinate and biternate flowers have been produced by such treatments. Fasciated flowers also occur occasionally, but there is no evidence that they occur more frequently than in other plants. In hyacinth, fasciation is encouraged by high temperature treatment, because more florets are borne on the thicker scape, giving a denser inflorescence. 'Tied-leaf' is a condition affecting up to 1–2% of some tulips. The demarcation between the last leaf and the flower is unclear, so the last leaf is petaloid or partly so in texture and colour and may be partly fused to the perianth. This interferes with the elongation of the last internode, bending the stem so the flower is held at an angle. Presumably the cause is a 'switching on' of flower initiation before the last leaf is complete, but there seems no clear link with any environmental or pathological condition.

Storage organs also suffer from some disorders which are believed to have non-parasitic causes, although secondary infections can soon occur. Such a case is that of soft rot of narcissus. Storage or transport of large quantities of recently lifted bulbs or those just given h.w.t. in poor ventilation, associated with high humidity, produce a rapid breakdown of the bulbs to a soft pulp with a grey-brown colour and a characteristic foetid smell. From data on bulb respiration, the heat generated by an insulated bulb mass is sufficient to provide composting conditions and a rapid breakdown.

The storage organs of several bulbs and corms can become dry and hard in storage, with scales that are white, opaque and chalk-like when broken. Later they become horn- or glass-like. Best known in tulip, the condition is known as 'chalking' (verkalking). It also occurs commonly in crocus. Usually, only a few 'bulbs' are affected, and it appears that mechanical or sun scald injury to the bulb at or just after lifting is the starting point, and exacerbated subsequently by poor ventilation and high humidity in store. Although several fungi have been isolated from affected bulbs, it seems unlikely that these are causal. The general advice is to improve bulb handling and storage techniques.

A disorder with some similarities to topple affects tulip bulb scales, producing brown necrotic spots especially on the outer fleshy scale. It appears that these result from the leakage of sap into the intercellular spaces. Plants of such bulbs are slow-growing, in proportion to the severity of the symptoms. It seems not to be related to any pathogen, and is not caused by micro-nutrient deficiency.

Gum formation in many 'bulbs' is promoted by ethylene generated from 'bulbs' infected with *Fusarium* (Fig. 9.3). Whilst this has a pathogenic cause, the 'bulbs' which produce the gum can themselves be free from disease. Further, other sources of ethylene (and to a lesser extent, other hydrocarbons) such as stored ripe fruit, boiler gases and CO_2 generating equipment can also induce gum production in healthy 'bulbs' of tulip, muscari, hyacinth, freesia and crocus. Gumming, and other disorders related to ethylene, can sometimes be observed in the field under waterlogged conditions, especially if associated with high temperatures before lifting. Gumming has been induced in healthy tulip bulbs by treatment with ethephon, with or without IAA. Analysis of the gum after hydrolysis showed a preponderance of xylose and arabinose with traces of glucose, mannose and uronic acid.

Hard base is a condition in tulip where root penetration through the base of the bulb is in some way impeded. It occurs commonly in Darwin hybrids lifted early before the outer scale has become dry and brown. The tunic is then tough and looks attractive for retail dry bulb sales. In five-degree forcing there is less time after planting for the tunic to disintegrate than in autumn planted bulbs, and the roots either grow in a bunch through the split tunic or are forced to grow upwards between the tunic and the outer fleshy scale. In such a situation, watering must be adequate and frequent to avoid losses of flowers.

Chemical damage occurs occasionally, and can be difficult to diagnose satisfactorily. Herbicide residues in the straw used for covering bulb boxes on standing grounds can distort emerging leaves and there is danger from spray drift from neighbouring crops. The use of desiccant sprays to speed leaf senescence before early lifting, and the application of herbicides after leaf die-down are both common practices, not without some danger to the 'bulbs' in the ground. It is wise practice to protect the bulbs by flailing off dead foliage and weeds before applying herbicide.

PLANT HEALTH INSPECTION AND CONTROL

Because bulb trade is international, and considerable quantities of bulbs are shipped all over the world, it is important that there are controls on the health of exported/imported stocks. Although most of the serious diseases of the commercially important crops are found in all the main bulb growing countries, present-day control measures can restrict or reduce further spread of pathogens, limit the spread of new pests and diseases and help to confine new and resistant strains.

All governments have a Plant Health Service to inspect imported plant material, to certify material prior to export, and to define health standards which must be met. For each export consignment, there must be an internationally recognized phytosanitary certificate signed by a plant health inspector, which confirms that all the requirements of the importing country have been complied with. This certificate accompanies the material, and it is an offence to send certifiable material to an EC member state without an appropriate health certificate. Many countries also require the exporter to have an import permit. Consignments are often inspected at the port of arrival, but much reliance is placed on the pre-export inspection done in the exporting country. Conditions are complex, with many countries insisting on complete absence of some pests, low tolerance of others and, sometimes, freedom from any adhering soil. Some consignments are required to be given disinfestation or disinfection before they are allowed into the receiving country. For some crops, a growing season inspection of the crop is also required, as some disease symptoms are apparent only on leaves and/or flowers.

About 80% of the 'bulbs' produced in The Netherlands, the world's largest producer, are exported, and this industry is supported by a statutorily based Bulb Inspection Service whose regulations are binding on every grower. The full-time staff of the Service, numbering several hundred, inspect crops growing in the field and 'bulbs' in store, at the exporters' shipping rooms, and at the port of despatch to ensure that product quality and freedom from disease are maintained to very high standards. Inspectors from the main importing countries are also stationed in The Netherlands. In other countries, grower funded certification schemes are operated to maintain standards of quality and ensure purchaser confidence in the product. There are also EC standards for flowering bulbs, corms and tubers.

10

FUTURE PROSPECTS

Prospects for the bulb industry are generally thought to be good, despite the constraints and differing requirements of a changing world. The 1980s have already seen many changes away from traditional bulb growing, and it is reasonable to expect that these changes, usually in response to financial considerations, and to purchaser opinion and pressure, will continue for some years.

Briefly, these changes are towards 'new' crops of many kinds, new products such as pot plants to replace cut flowers, improved techniques for treating, handling and forcing 'bulbs', changes in husbandry to protect the environment and the need to reduce the use of pesticides. There are also indications of increased effort to educate the buyers of bulbs in how to grow them to best advantage, and what to grow where, and the improvement and maintenance of product quality is a continuing challenge. To support these changes there is continuing need for research input, so that changes implemented have a sound scientific base, and a need to maintain close links between the production and marketing sectors of the industry, the advisory and plant health services and research and development.

NEW PRODUCTS

Novelty is an important component of selling flowers, so in this sense the production of new cultivars of well-known species has always been part of 'bulb' growing. The recent increasing interest in morphologically different cultivars, especially dwarf ones such as narcissus Tête-a-Tête, and genetically short tulips, has led to an increased demand for pot plants because of their longer life in the modern heated home than the more traditional cut flowers. Pot plants also represent a better financial return for the grower, albeit at a greater cost of production and difficulties of transport to market. In Europe there is a tendency to more pot-plant production, rather than cut flowers, a situation that has existed for a long time in the USA.

But there is more to pot-plant production than growing successful and popular cut-flower cultivars in pots. The aesthetic appearance of the final product has to appeal to the purchaser, particularly the height of the leaves and flowers. Height can be controlled to differing degrees by manipulating the temperatures pre- and post-planting and the duration of the cold treatment, by chemical means, or by starting with a genetically short (but usually expensive) cultivar. For success it is necessary to choose the best alternative for the species and cultivar being grown to avoid some of the known side-effects and produce a well-defined quality product for the lowest cost.

In addition to the already well-known genera and species, there are many others, new to horticulture, or whose potential for becoming economically important horticultural crops has not hitherto been realized. There are rich floras waiting to be tapped: of the 20 000 indigenous plant species in South Africa, c. 2700, in 15 families, are 'bulbous'. Similarly, there are many potential crop plants in South America. At Lisse a collection of about 100 genera comprising 400 species, types or hybrids has been assembled for assessment, from Australia, Chile, Colombia, Lesotho, Malawi and the USA (Koster, 1989). It is a long and slow process to develop a wild species into a commercially valuable crop plant. Paradoxically, 'new' flower crops are not usually completely unknown; most are minor horticultural plants which have been 'improved' to make them suitable horticultural subjects.

There are three main areas of progress: in the first, existing selections and clones which are already well established crop plants are improved by further fine tuning to produce new cultivars. This is the major activity of many commercial breeders, producing new cultivars of narcissus, tulip, iris and lily which have features making them more acceptable to the grower or the buying public. These features are of two kinds, the first being a characteristic such as improved disease resistance, a shorter low-temperature requirement or improved yield potential (features that improve the profitability of crop husbandry) or have morphological or physiological characteristics that appeal to the purchaser of the 'bulbs' (increased hardiness, dwarf stature) or the flowers (better flower form and/or colour, longer vase life).

In the second, a new horticultural use is perceived for an existing, and often well-known species, such as the development of a garden ornamental into a pot plant. A well-known past example here is *Alstroemeria*, grown in British gardens for nearly two centuries as *A. ligtu* and *A. aurantia* hybrids, but now hybridized to give a range of cultivars with highly attractive, long lasting blooms, grown so widely under glass that it is in the top ten of Dutch cut-flower species.

The third requires research on little known species with potential, for which little information on cultural methods exists. This is the most interesting group because of the possibility of a novel final product. It is also the most difficult, because it requires vision to identify potential, initial studies to establish basic growing methods, followed by detailed assessments by experienced growers of production schedules and marketing strategy. Only then will it be possible to decide whether it has the makings of a major crop plant. During preliminary evaluation, it is necessary to build up a crop profile to assess a number of factors such as flower appearance, numbers, duration of flowering, keeping quality, plant habit, environmental requirements, pest and disease resistance and an estimate of market acceptability. If the plant looks promising, shortcomings can be investigated with a view to breeding improved types. Even after the new product is on the market, consumer opinion needs to be sought to introduce further improvements (such as widening the colour range) to counter the flagging interest which always follows some time after the introduction of a successful new product.

Once a new plant has been screened to the stage of introduction to the market, rapid propagation methods are available to ensure that there are sufficient supplies to make a big initial impression, which can be helped by advertising and publicity in the trade press.

New Pot Plants

The following species have been mentioned in recent literature as possible pot-plant subjects, usually small bulbs with a maximum height of 20–25 cm for the European market, although Americans prefer taller ones: *Achimenes, Anemone blanda, Fritillaria meleagris, Oxalis adenophylla, Rhodohypoxis baurii, Tulipa greigii* and *T. kaufmanniana*. Tall and genetically short cultivars of lily have proved successful as pot plants when dwarfed with daminozide, uniconazole or ancymidol, several cultivars achieving the ideal height for the USA market of 30–46 cm. A range of freesia cultivars treated with ancymidol made promising pot plants, whilst a dwarf cultivar of *Zantedeschia aethiopica* showed possibilities for year-round production as a pot plant, although some chemical treatment might be necessary to keep the plants short in low winter light. The narcissus cv. Tête-a-Tête has become an established pot plant because of its small size (15–20 cm) and relatively large flowers.

New Species, Hybrids and Selections

New cultivars of gladiolus have been evaluated for their suitability for early glasshouse flowering, and trials of a range of narcissus cultivars have sought to extend the current restricted number used in forcing for cut flowers in The Netherlands. *Zantedeschia elliottiana* and *Z. rehmannii* cultivars are being assessed for their suitability for growing outdoors and under glass. Trials of cultural methods for *Tritelia laxa* have been done in California, and there is increasing interest in *Liatris* and *Polianthes* in several countries.

Some potentially useful new cultivars or species could be developed as pot plants or as cut flowers, and an evaluation in The Netherlands of 11 cultivars of hippeastrum dealt with both. The first few new bulb crops assessed at Lisse were released for grower trial in 1989 as cut flowers, garden plants or as pot plants. The following genera were included: *Agapanthus*, *Amaryllis*, *Polygonatum*, *Scadoxus* and *Tulbaghia*. Other genera currently of interest are *Ornithogalum* spp. as rock plants, polyploid *Sparaxis* as cut flowers and *Achimenantha*, a hybrid between *Achimenes* and *Smithiana*.

IMPROVED TECHNIQUES

The present trend to improve methods of handling 'bulbs' looks set to continue. The industry is already well mechanized for planting, lifting, grading, cleaning and storing 'bulbs', especially on the larger production units where economies of scale mean that they are in a better position to operate economically than the smaller holdings, which are consequently becoming fewer. As much of the machinery on 'bulb' farms is in use for only a short time each year, with little possibility of it being used for other crops at other times (despite some compatibility with onions, potatoes and other vegetables), it represents a disproportionately large capital investment which makes it difficult for the small-scale grower. Similarly small stores are more expensive to construct and run, on a unit volume basis, than large ones. The tendency is therefore to larger, highly mechanized specialist growers. Whilst mechanical handling and general mechanization is expensive in terms of equipment and of the energy required to run it, it is preferable to manual methods because of the time saved, the greater flexibility and capability and the high cost of labour itself.

For bulb forcing, also, economies of scale apply, with greater mechanization of handling 'bulb' trays into and out of stores and glasshouses. Modern glasshouses, well laid out in relation to other buildings, and equipped with efficient heating systems, supplementary and photoperiodic lighting (where appropriate), computerized control of temperature and the use of thermal screens mean that the energy input into forcing bulbs is kept to a minimum, production costs are kept down, and the flowers can then be sold at competitive prices.

Better information continues to be made available on all aspects of crop growing and flower production. On field growing, there are models of optimum planting densities for the major crops to suit requirements for different bulb grades. Whilst these are not widely used yet, because of the seasonal variations in weather which make fine predictions difficult, they are still a big improvement on the highly empirical methods of the past. Further improvements can be expected by incorporating weather probabilities, and more detailed information on the responses of individual cultivars, but this will require considerable and continued research input.

Details are now available on the cold requirements of established cultivars, and are being produced as new ones become available; accurate scheduling of successive crops can now be used to meet marketing dates and ensure continuity of supplies. This depends on the good control of the forcing house environment by improved systems allied to efficient planning and management. But as new cultivars and species are developed, a great deal of work will be required to develop efficient methods of storage, 'bulb' treatments, growing systems, and flower harvesting and storage. For newer cultivars of well established 'bulb' crops, this will be a fine tuning exercise, but 'new' species will require considerable input.

Little progress has so far been made in the use of genetic engineering for improving ornamental 'bulbs', but the potential exists for transferring into current cultivars desirable characteristics as diverse as disease resistance, flower colour, duration of cold requirement and plant height. Such changes would not affect other aspects of the genotype that would necessitate a full re-investigation of cultural methods as is the case at present when a new cultivar is bred.

Propagation methods have been improved considerably in the past decade, but commercial uptake of specialized methods such as micropropagation has been modest for some species although encouraging for others. The high cost of micropropagation makes it of only marginal interest in cheaper 'bulbs' where natural means and the simpler artificial methods are perfectly adequate except for special circumstances. In The Netherlands, which tends to lead in the adoption of new techniques, micropropagation of lilies is common, with over 12 million plants being produced annually, compared with only about 200 000 narcissus. For special cases, the rapid *in vitro* techniques available are being employed for bulking up new sports or cultivars, virus-tested individual plants and in the embryo rescue of difficult crosses. Similar methods will probably be developed for use with other genera, like tulip, whose micropropagation has hitherto proved most intractable.

In general, the commercial use of plant growth regulators (PGRs) has not been as great as had been anticipated some years ago, despite the demonstration that endogenous PGRs are important in the control of plant processes like stem extension, the onset and breaking of 'dormancy', flower initiation and the post-initiation development of the flower to anthesis. Several compounds are being used successfully as dwarfing compounds to shorten plants for pot growing and to prevent excessive height of cut-flower crops, and ethephon will probably continue to be used for overcoming dormancy and improving flowering of tazetta narcissus, freesia and iris. If and when the responses of the different species to the range of PGRs is better understood, and better methods are available for applying the chemicals to their sites of action, there are great possibilities for modifying plant behaviour in efficient production systems, including *in vitro* growing.

MARKETING

The scope for increased sales of 'bulbs' and 'bulb' flowers is indicated by the great disparity in flower numbers bought in different countries, with the USA and UK being poorer consumers on a *per capita* basis than other European countries, especially The Netherlands. 'Bulbs' are a seasonal product, as are most 'bulb' flowers, so that promotional effort must be mounted every year to renew purchaser interest. At the same time, new products can be added to the available range, with information on how and when to plant 'bulbs' and how to maximize the flower vase life or the duration of flowering of a pot plant.

As there are three components to the purchasing sector – the forcing industry, the landscape users and the household users – appropriate approaches must be made to maintain and increase sales. The forcer must have the backing of the bulb producer to take up new products, must be aware of new techniques and have a modern, entrepreneurial attitude to his/her business. Distribution and marketing of his/her products must be efficient and aggressive to improve sales. Those concerned with landscaping need to be informed of suitabilities of species for different locations, need to be aware of planting techniques, and realize the potential for using perennial (naturalized) plantings. For the domestic market, the aim must be to increase the numbers of purchasing households and the numbers bought by each. To this end, retailers need to be informed of the possibilities of 'bulb' and flower sales, and must promote these products to the purchaser. It is relevant that a recent survey of trends in public purchase of gardening products in the UK showed that 'neatly packaged, convenient products, e.g. bulbs', was the sector growing fastest – an indication of the importance of presentation for sales.

In the promotion of the sale of 'bulbs' and flowers, there is a need for a central body to coordinate information, to distribute advertising material and to supply technical information to the trade. Such a role is played by various organizations in The Netherlands associated with the International Flower Bulb Centre at Hillegom, which has links with national centres like the North American Flower Bulb Wholesalers and the Bulb Information Desk in the UK. Within the UK, there is a Bulb Distributors' Association based at Spalding in Lincolnshire.

A great deal of publicity is generated by Springfields Gardens at Spalding, supported by British growers and showing the range of British 'bulbs', and, of course, the Keukenhof gardens in The Netherlands, both of which are tourist attractions as well as demonstrating bulb products to the public and trade. Trade interests in 'bulb' growing in the UK are catered for by the

Springfields Horticultural Exhibition (SHE) held at Spalding annually in the first week of February.

CROP PROTECTION

Several factors combine to make crop protection an important issue. The effectiveness of currently available chemicals is decreased by the resistance developed to them by the pests and disease organisms, and, at the same time, many chemicals are being withdrawn from the pathologist's arsenal because they are no longer approved for environmental or health reasons. Fewer new chemicals are coming onto the market because costs of developing such chemicals are increasing. In the past, pesticides have been important in allowing the 'bulb' industry to develop and expand to its present size and have permitted the setting of high standards of product quality. In future, there must be less reliance on chemicals for such purposes as soil sterilization, weed control and for treating crops affected by pests and diseases. There is a strong, but ill-informed, public lobby against the use of chemicals which are regarded as socially unacceptable. Whilst this feeling is most strongly directed against food crops, growers of ornamentals also suffer because the reduced market for their products dissuades chemical manufacturers from developing new products. UK users have to comply with the requirements of the Food and Environment Protection Act (1985) and the Control of Pesticides Regulations (1986).

A government scheme in The Netherlands aims to reduce by 60% by the end of the decade the present rate of use of insecticides and fungicides ($c.$ $4.5\,kg\,ha^{-1}$ annum $^{-1}$) by bulb growers. This official concern stems from the toxic effects of the chemicals on workers in the industry and increasing pollution of ground water.

Newer methods are being sought to maintain crop health and product quality (van Aartrijk, 1990). Currently under consideration in The Netherlands are the production of initially healthy stocks based on meristem-tip culture and temperature treatments to eliminate pathogens, allied to rapid vegetative propagation methods. The use of pathogen resistant genotypes will depend on developing resistance tests for screening existing and new genotypes, and the breeding of new resistant types, again allied to rapid propagation methods for bulking successful forms. Cultivation techniques can be modified to grow valuable material in more sterile conditions, and field material can be grown under longer rotations. More use is envisaged of physical control measures such as steam sterilization, and of biological control methods, especially in 'closed' situations such as stores and glasshouses. Some chemical control will still be permitted, as far as can be foreseen, but quantities and frequencies will be kept to a minimum, with less adverse impact on the environment by reducing spray drift and spillage. By integrating these methods into cultural systems it is hoped that the crop protection aspect of 'bulb' growing will be revolutionized with little loss of volume or quality. It is certainly a challenge to the growers, extension service and research workers.

INDUSTRIAL BACK-UP

The viability of the 'bulb' industry, nationally and internationally, depends heavily on close contact and cooperation between many components: marketing and sales staff, exporters, the growers, advisory and extension staff, plant health officers and those concerned with research and development. By exchange of information through these groups, the preferences and complaints of the consumer get through to those who can do something about them, and in the reverse direction, the growers and marketing organizations find out what is possible and feasible, as well as what is being investigated, with both short- and long-term goals. The advisory and plant health services are aware of the extent and severity of the problems of the moment, and can either provide quick answers from known remedies, or refer new or intractable problems to the research workers. By this cooperation, a range of desirable 'bulb' products of high standards of health and quality can be produced in an efficient way at affordable prices, and the 'bulb' industry will flourish. In the Preface I stressed the vast amount of information already available and continuing to be added to existing knowledge. It is already far beyond human capacity to retain all that is known about 'bulbs' even in a single discipline, and a readily accessible database is becoming essential for anyone professionally involved with the crops. The forthcoming comprehensive treatise *The Physiology of Flower Bulbs* edited by De Hertogh and Le Nard will go some way to satisfying this need.

REFERENCES

Alford, D.V. (1991) *Pests of Ornamental Trees, Shrubs and Flowers*. Wolfe Publishing Ltd (Mosby-Year Book Europe), London.

Baardse, A.A. (1977) *Groot Lelie Boek*. Drukkerij 'West Friesland' bv, Hoorn, The Netherlands.

Balls, R.C. (1985) *Horticultural Engineering Technology – Field Machinery*. Macmillan, London.

Balls, R.C. (1986) *Horticultural Engineering Technology – Fixed Equipment and Buildings*. Macmillan, London.

Benschop, M. (1980) Photosynthesis and respiration of *Tulipa* sp. cultivar 'Apeldoorn'. *Scientia Horticulturae* 12, 361–375.

Blanchard, J.W. (1990) *Narcissus: A Guide to Wild Daffodils*. Alpine Garden Society, Woking, Surrey, UK.

Brunt, A.A., Price, D. and Rees, A.R. (Revisers) (1979) Moore, W.C., *Diseases of Bulbs*, 2nd edn. MAFF Reference Book HPD 1. HMSO, London.

Cremer, M.C., Beijer, J.J. and de Munk, W.J. (1974) Developmental stages of flower formation in tulips, narcissi, irises, hyacinths and lilies. *Mededelingen Land-bouwhogeschool, Wageningen* 74, 15.

Dafni, A., Cohen, D. and Noy-Meir, I. (1981) Life cycle variation in geophytes. *Annals of the Missouri Botanic Gardens* 68, 652–660.

Dahlgren, R.M.T. and Clifford, H.T. (1982) *The Monocotyledons: A Comparative Study*. Academic Press, London.

De Hertogh, A. (1989) *Holland Bulb Forcer's Guide*, 4th edn. The International Flower Bulb Centre, Hillegom, The Netherlands.

De Munk, W.J. (1989) Thermomorphogenesis in bulbous plants. *Herbertia* 45, 50–55.

Dicks, J.W. and Rees, A.R. (1973) Effects of growth-regulating chemicals on two cultivars of Mid-Century hybrid lily. *Scientia Horticulturae* 1, 133–142.

Doerflinger, F. (1982) *Know Your Bulbs. Manual No. 1 'Spring Flowering Bulbs'*. Horticultural Trades Association, Reading, UK.

Doerflinger, F. (1983) *Know Your Bulbs. Manual No. 2 'Summer Flowering Bulbs'*. Horticultural Trades Association, Reading, UK.

Doorenbos, J. (1954) Notes on the history of bulb breeding in the Netherlands. *Euphytica* 3, 1–11.

Doss, R.P. (1981) Effects of duration of 32°C and 20°C postharvest bulb treatments on early forcing of Ideal iris. *Canadian Journal of Plant Science* 61, 647–652.

Everett, T.H. (1954) *The American Gardener's Book of Bulbs*. Random House, New York.

Fernandes, A. (1967) Keys to the identification of native and naturalized taxa of the genus *Narcissus*. *RHS Daffodil and Tulip Yearbook 1968* 33, 37–66.

Fry, B.M. (1978) Progress of the narcissus breeding programme. *Report of Rosewarne Experimental Horticulture Station for 1977* 20–28.

Galil, J. (1961) *Kinetics of Geophytes*. Hakibbuts Hameuchad Ltd, Tel-Aviv.

Gilford, J. McD. and Rees, A.R. (1974) The tulip shoot apex. I. Structure and development. *Scientia Horticulturae* 2, 1–10.

Goemans, R.A. (1980) The history of the modern freesia. *Linnaean Society Symposium Series* No. 8, Petaloid Monocotyledons, 161–170.

Gratwick, M. and Southey, J.F. (eds) (1986) *Hot-water Treatment of Plant Material*. MAFF Reference Book 201. HMSO, London.

Halevy, A.H. (1990) Recent advances in the control of flowering and growth habit of geophytes. *Acta Horticulturae* 266, 35–42.

Hanks, G.R. (1982) The response of tulips to gibberellins following different durations of cold storage. *Journal of Horticultural Science* 57, 109–119.

Hanks, G.R. and Jones, S.K. (1986) Notes on the propagation of *Narcissus* by twin-scaling. *The Plantsman* 8, 118–127.

Hanks, G.R. and Rees, A.R. (1977) Stem elongation in tulip and narcissus: the influence of floral organs and growth regulators. *New Phytologist* 78, 579–591.

Hanks, G.R. and Rees, A.R. (1979) Photoperiod and tulip growth. *Journal of Horticultural Science* 54, 39–46.

Hanks, G.R. and Rees, A.R. (1984) Early forcing of narcissus: the effects of lifting date and stage of floral development at the start of cooling. *Scientia Horticulturae* 23, 269–278.

Ho, L.C. and Rees, A.R. (1975) Aspects of translocation of carbon in the tulip. *New Phytologist* 74, 421–428.

Ho, L.C. and Rees, A.R. (1976) Re-mobilization and redistribution of reserves in the tulip bulb in relation to new growth until anthesis. *New Phytologist* 76, 59–68.

Ho, L.C. and Rees, A.R. (1977) The contribution of current photosynthesis to growth and development of the tulip during flowering. *New Phytologist* 78, 65–70.

Holttum, R.E. (1955) Growth habits of monocotyledons, variations on a theme. *Phytomorphology* 5, 399–413.

Hussey, G. (1978) The application of tissue culture to the vegetative propagation of plants. *Science Progress, Oxford* 65, 185–208.

Imanishi, H. (1983) Effects of exposure of bulbs to smoke and ethylene on flowering of *Narcissus tazetta* cultivar 'Grand Soleil d'Or'. *Scientia Horticulturae* 21, 173–180.

Kamerbeek, G.A., Beijersbergen, J.C.M. and Schenk, P.K. (1970) Dormancy in bulbs and corms. *Proceedings of the 18th International Horticultural Congress* 5, 233–239.

Koster, J. (1989) Introduction of new bulbous crops in The Netherlands. *Herbertia* 45, 30–32.

Lane, A. (1984) *Bulb Pests*. MAFF Reference Book 51. HMSO, London.

Lang, G.A. (1987) Dormancy; a new universal terminology. *HortScience* 22, 817–820.

Lawson, H.M. and Wiseman, J.S. (1978) The effect of weeds on the growth and development of narcissus. *Journal of Applied Ecology* 15, 257–272.

Le Nard, M. (1983) Physiology and storage of bulbs: concepts and nature of dormancy in bulbs. In: Lieberman, M. (ed.), *Post Harvest Physiology and Crop Preservation*. Plenum Press, New York, pp. 191–230.

Linfield, C. and Lole, L. (1989) Pests and diseases of outdoor bulbs and corms. In: Scopes, N. and Stables, L. (eds), *Pest and Disease Control Handbook*, 3rd Edn. BCPC, Bracknell, UK, Chapter 14, pp. 603–617.

MAFF (1976) *Narcissus Forcing*. Booklet B2299. HMSO, London.

MAFF (1977) *Anemones*. Advisory Leaflet 353. HMSO, London.

MAFF (1981) *Tulip Forcing*. Booklet B2300. HMSO, London.

MAFF (1982a) *Tulip Bulb Production*. Booklet B2298. HMSO, London.

MAFF (1982b) *Earlier Outdoor Narcissus Production*. Booklet B2398. HMSO, London.

MAFF (1984a) *Lilies*. Leaflet 859. HMSO, London.

MAFF (1984b) *Bulb and Corm Production*, 5th edn. Reference Book 62. HMSO, London.

MAFF (1985) *Narcissus Bulb Production*. Booklet B2150 (revised). HMSO, London.

Mathew, B. (1978) *The Larger Bulbs*. Batsford, London.

Mathew, B. (1982) *The Crocus*. Batsford, London.

Mathew, B. (1987a) *The Smaller Bulbs*. Batsford, London.

Mathew, B. (1987b) *Flowering Bulbs for the Garden*. Collingridge, London.

Nowak, J. and Rudnicki, R.M. (1990) *Postharvest Handling and Storage of Cut Flowers, Florist Greens and Potted Plants*. Chapman and Hall, London.

Price, D. (1970) Tulip fire caused by *Botrytis tulipae* (Lib.) Lind.; the leaf spotting phase. *Journal of Horticultural Science* 45, 233–238.

Raunkier, C. (1934) *The Life Forms of Plants and Statistical Plant Geography*. Clarendon Press, Oxford.

Read, M. (1989) Over-exploitation of wild bulbs by the horticultural trade. *Herbertia* 45, 6–12.

Rees, A.R. (1966) Dry-matter production by field-grown tulips. *Journal of Horticultural Science* 41, 19–30.

Rees, A.R. (1967) The crop respiration rate of tulips. *New Phytologist* 66, 251–254.

Rees, A.R. (1968) The initiation and growth of tulip bulbs. *Annals of Botany* 32, 69–77.

Rees, A.R. (1969) The initiation and growth of *Narcissus* bulbs. *Annals of Botany* 33, 277–288.

Rees, A.R. (1972) *The Growth of Bulbs*. Academic Press, London.

Rees, A.R. (1977) The cold requirement of tulip cultivars. *Scientia Horticulturae* 7, 383–389.

Rees, A.R. (1985) Ornamental bulbous plants. In: Halevy, A.H. (ed.), *Handbook of Flowering Vol. 1*. CRC Press, Boca Raton, Florida, pp. 259–308.

Rees, A.R. (1986) Narcissus: flowers per tonne of bulbs. *Acta Horticulturae* 117, 261–266.

Rees, A.R. (1987) Forcing hyacinths. *The Garden. Journal of the Royal Horticultural Society* 112, 417–420.

Rees, A.R. and Briggs, J.B. (1974) Optimum planting densities for tulips grown in ridges in the field. *Journal of Horticultural Science* 49, 143–154.

Rees, A.R. and Thornley, J.H.M. (1973) A simulation of tulip growth in the field. *Annals of Botany* 37, 121–131.

Rees, A.R. and Turquand, E.D. (1969) Effects of planting density on bulb yield in the tulip. *Journal of Applied Ecology* 6, 349–358.

Rees, A.R., Wallis, L.W., Turquand, E.D. and Briggs, J.B. (1972) Storage treatments for early flowering of narcissus. *Experimental Horticulture* 23, 64–71.

Roberts, A.N., Stang, J.R., Wang, Y.T., McCorkle, W.R., Riddle, L.L. and Moeller, F.W. (1985) *Easter Lily Growth and Development*. Technical Bulletin 148, Oregon State University, Corvallis, Oregon, USA.

Roh, M.S. (1989) Control of flowering in *Lilium* – a review. *Herbertia* 45, 65–69.

Schmida, A. and Dafni, A. (1989) Blooming strategies, flower size and advertising in the "lily-group" geophytes in Israel. *Herbertia* 45, 111–123.

Scopes, N. and Stables, L. (Eds) (1989) *Pest and Disease Control Handbook*. BCPC, Thornton Heath, UK.

Smith, D. (1979) *Freesias*. Grower Guide No. 1. Grower Books, London.

Squires, W.M. and Langton, F.A.L. (1990) Potential and limitations of narcissus micropropagation: an experimental evaluation. *Acta Horticulturae* 266, 67–75.

Van Aartrijk, J. (1990) Changing views on chemical crop protection in the Netherlands and the consequences for bulb research. *Acta Horticulturae* 266, 385–389.

Van der Valk, G.G.M. and Timmer, M.J.G. (1974) Plant density in relation to tulip bulb growth. *Scientia Horticulturae* 2, 69–81.

Van der Valk, G.G.M. and Van Gils, J.B.H.M. (1990) Structure and applications of a production model in tulip bulb culture. *Acta Horticulturae* 266, 391–396.

Van Eijk, J.P., Nieuwhof, M., Van Keulen, H.A. and Keijzer, P. (1987) Flower colour analyses in tulip (*Tulipa* L.). The occurrence of carotenoids and flavonoids in tulip tepals. *Euphytica* 36, 855–862.

Wang, Y.T. (1990) Growth and leaf photosynthesis of *Lilium longiflorum* 'Nellie White' in response to partial defoliation after anthesis. *Acta Horticulturae* 266, 197–204.

Warrington, I.J., Seager, N.G. and Armitage, A.M. (1989) Environmental requirements for flowering and bulb growth in *Nerine sarniensis*. *Herbertia* 45, 74–80.

Webb, D.A. (1980) Narcissus. In: Tutin, T.G., Heywood, V.H., Burges, N.A., Moore, D.M., Valentine, D.H., Walters, S.M. and Webb, D.A. (eds), *Flora Europaea Vol. 5*. Cambridge University Press, pp. 78–84.

Widmer, R.E. (1980) Cyclamen. In: Larsen, R.A. (ed.), *Introduction to Floriculture*. Academic Press, London, pp. 375–394.

Woodcock, H.B.D. and Stearn, W.T. (1950) *Lilies of the World*. Country Life Ltd, London.

Yasui, K., Miyata, K. and Konishi, K. (1974) Histological studies on formation and thickening growth of gladiolus corms. *Journal of the Japanese Society for Horticultural Science* 42, 371–379.

INDEX